DENIAL

Self-Deception, False Beliefs,
and the Origins of the Human Mind

AJIT VARKI

AND

DANNY BROWER

TWELVE

New York Boston

Twelve
Hachette Book Group
237 Park Avenue
New York, NY 10017

www.HachetteBookGroup.com

Printed in the United States of America

RRD-C

First Edition: June 2013
10 9 8 7 6 5 4 3 2 1

Twelve is an imprint of Grand Central Publishing.
The Twelve name and logo are trademarks of Hachette Book
Group, Inc.

The Hachette Speakers Bureau provides a wide range of authors
for speaking events. To find out more, go to
www.hachettespeakersbureau.com or call (866) 376-6591.

The publisher is not responsible for websites (or their content)
that are not owned by the publisher.

Library of Congress Control Number: 2013932348

For Pothan Joseph and Anna Varki,
who bequeathed both nature and nurture to me
in great measure,

and for

the late Danny Brower,
without whom this book would not have existed

CONTENTS

DENIAL

DE·NI·AL

An unconscious defense mechanism characterized by refusal to acknowledge painful realities,
thoughts, or feelings.

*—The American Heritage Dictionary
of the English Language*

We now know that the human animal is characterized
by two great fears that other animals are protected from:
the fear of life and the fear of death.

—Ernest Becker, in *The Denial of Death*

We who perhaps one day shall die, proclaim man as immortal at the flaming heart of the instant.

—Saint-John Perse (Alexis Saint-Léger Léger),
in *Seamarks*

The yaksha asked: "What is the greatest surprise?"
Yudhisthira replied: "People die every day, making us
aware that men are mortal. Yet we live, work, play, plan,
etc., as if assuming we are immortal. What is more surprising than that?"

—The Mahabharata

An Improbable but True Story

Truth is stranger than fiction.

—Lord Byron, in *Don Juan*

Fiction is obliged to stick to possibilities; truth isn't.

—Mark Twain, in *Following the Equator*

The story behind this book is strange and improbable. Two individuals from very different backgrounds converged on a single question, happened to meet just once, discussed it briefly, parted company—and would never see each other again. One of them, Danny Brower, died suddenly at the age of fifty-five in 2007. The other person—I, a physician turned scientist—was left to complete our story. From our single chance conversation grew this book, which should interest anyone who cares about the universally human questions *Who are we? How did we get here? Why are we the way we are? And where are we going?*

The improbability of it all becomes starker when you consider what different circumstances the two of us came from. Danny was born in November of 1951, was raised in the United States, and worked his way up from modest means to the prestigious position of professor and chair of molecular and cellular biology at the University of Arizona at Tucson. By the time I met him, he was already well known for his pioneering work on protein molecules called integrins,[1] which play key roles in how cells recognize and respond to their environment. Danny was using a popular fruit-fly model to study these processes, and from this same work, he was even able to contribute to our understanding of human cancer. As it turned out, Danny had another interest related to his research—he was fascinated by evolutionary biology,[2] the study of the processes by which all life on this planet emerged over the last three billion years or so. A natural progression of such thinking made him wonder about the origin of our own species, *Homo sapiens*.

As for me, I was born just two months after Danny, but was raised on the other side of the planet, in India. I grew up in a traditional Orthodox Christian family from the southern state of Kerala, but attended English-language schools and went on to medical college with the idealistic goal of saving lives. But as it happens, the curriculum in medicine includes a strong dose of fascinating biology. Inspired by this aspect of my education, I finally decided that I could contribute more to society by becoming a biomedical researcher. However, opportunities to pursue this track in India were sparse in the 1970s. Reading the scientific literature, I realized that the United States was the one country in the world where physicians were being encouraged and supported in their efforts to do research side by side with other kinds of scientists. Thus it was that I emigrated to the United States in 1975 with six dollars and a suitcase, eventually becoming board certified in internal medicine, hematology, and oncology and working my way up to my present position as a professor at the University of California, San Diego (UC San Diego). Just as I had originally hoped, this career path allowed me to pursue my passion for science and research, and eventually took me away from patient care and into the emerging field of glycobiology, which studies the dense, complex, and varied forest of sugar chains that are now known to cover every one of the cells in our bodies.[3] Technical difficulties in analyzing these "glycans" resulted in their getting limited attention in the early stages of the molecular biology revolution, which had focused mostly on DNA, RNA, and proteins. But we now know that these glycan chains are essential for life, and that they are involved in every normal and abnormal state of the body, from infections to cancer to brain development.[4]

While starting up my independent research career I still continued to see patients part-time, as a physician and cancer specialist. The latter role naturally led me to ponder issues of life and death, particularly the question of how it was that patients with terminal cancer could so courageously fight to stay alive against all odds. It seemed to me that both patient and physician were actually denying the reality of what they were up against, even in the face of a grim prognosis. But then, optimistic thinking that helps us go on despite the odds doesn't just feature in life-or-death situations; it is part and parcel of what makes us human, and comes across in so many of our activities. These and other life experiences, such as watching my own daughter grow up,[5] made me wonder about how we became human, evolving away from a recent common ancestor with our closest living evolutionary cousins, the chimpanzees and other so-called great apes— gorillas, bonobos, and orangutans.[6] Although great apes and humans look rather different, scientists as far back as in the 1960s and 1970s had shown that we are genetically very similar. In fact, viewed from the perspective of genes, we are more similar to chimpanzees than mice and rats are to each other! And chimpanzees have more in common with us genetically than they do with gorillas. So the big question has been: Why are we humans so different from chimpanzees and gorillas in appearance, behavior, and so many other features while they seem so similar to each other? Why is it that a chimpanzee or gorilla cannot do what I am doing right now— communicating with a reader about stories of past events with implications for our future? And although we may have never met each other, how is it that you, the reader, understand what I am thinking, and how do I know that you might be doing so?

In 1984, my thoughts about such matters were very suddenly brought into focus. I was seeing a patient who had an immune reaction to a horse serum that had been administered to treat a rare blood disease. What I learned from this case inspired additional research, and by the mid-1990s my research group had uncovered the first known clear-cut genetic difference between humans and great apes.[7] In fact, scientists had been searching for two decades for genetic features that were uniquely different in humans, and I was lucky enough to find the very first example, the loss of a gene called *CMAH*, which had subtly but uniquely altered the cell surface sugars called sialic acids on all cells in the human body. Since then we have found several additional uniquely human changes in sialic acid biology that seem to contribute to the human condition, in health and disease.[8] But that's another story, for another time.

These unexpected findings stoked my already keen interest in something quite far removed from my original training—an explanation for the origins of the human species. Where we humans came from is undoubtedly one of our greatest unsolved mysteries, at least from the human perspective. And while the work of many scientists had painted the broad brushstrokes of how this might have happened, there was precious little known at the time about any molecules and biological processes unique to humans. So by the late 1990s I began to focus my research specifically on this area of anthropogeny (this classic but long-unused term encompasses the scientific pursuit of human origins and evolution).[9] As you can imagine, this fascinating field of inquiry requires understanding of a rather wide range of subjects, many of which I was not trained in. But having achieved a modicum of scientific success and recognition in my primary research fields of medicine and glycobiology, I could now afford to devote more time to this new quest, immersing myself in other relevant specialties, such as primatology, evolutionary biology, neuroscience, linguistics, and anthropology. As part of this self-education quest, I began to seek advice from various experts in such fields, a process that eventually led to my forming (in 1996) an international transdisciplinary collaborative group of researchers interested in human origins and evolution.[10] Supported primarily by the Mathers Foundation of New York,[11] this group was recently expanded and renamed the Center for Academic Research and Training in Anthropogeny (CARTA).[12] In this "center without walls," affiliated with UC San Diego and the Salk Institute for Biological Studies, CARTA brings together academicians from the natural sciences, social sciences, and biomedical sciences along with interested parties from the humanities and arts, as well as contributors from the earth sciences, engineering, mathematics, and computing sciences. This first-of-its-kind effort assumes that definitive answers about the human origins mystery can best arise from breaking down traditional academic barriers and drawing experts from every relevant discipline into a creative transdisciplinary discussion. The core mission of CARTA is to *"use all rational and ethical approaches to seek all verifiable facts from all relevant disciplines to explore and explain the origins of the human phenomenon."*[13]

But let's get back to the story of how I met Danny Brower. A decade into my quixotic quest to understand human origins and evolution, my own knowledge base was sufficient to embolden me to give a few public lectures on the topic. One of the first I delivered was at the University of Arizona on April 2, 2005, about molecular differences between humans and chimpanzees and how they might have contributed to human uniqueness. As you might imagine, I was a bit nervous. But the lecture seemed to go well, and audience responses were positive. At the pleasant spring open-air luncheon that followed, a tall, intense man with a scraggly beard sat down next to me, introduced himself as Danny Brower, and pointedly informed me that we were "all asking the wrong question." At first I thought he was some local eccentric, but when I realized he was a well-known professor at the university, I gave him a careful hearing.

Instead of just asking what evolutionary processes made us human, Danny said we should also be asking why such complex mental abilities have appeared *only* in humans, despite many other intelligent species having existed and evolved for millions of years. In other words, if having complex human-like mental abilities has been so good for the success of our species (as everyone has assumed), then how is it that we are the only species that got so brainy? The usual assumption is that something very unusual and special happened to human brains during evolution, and that we just need to find out what that something is. But Danny took a fascinating contrarian's position, saying that we should *not* be looking for what everyone else was—the presumed special brain changes that made us human. Rather, we should be asking *what has been holding back* all the other intelligent species that, like humans, seem to have self-awareness of themselves as individuals[14]—a list that may include chimpanzees, orangutans, dolphins, orcas (so-called killer whales), elephants, and even birds such as magpies. Danny asked: Why are there no humanlike elephant or humanlike dolphin species as yet, despite millions of years of evolutionary opportunity for making this transition?

The next mental step beyond the basic awareness of one's own personhood that many of the species mentioned above seem to possess could be awareness of the personhood of others—in other words, knowing that others of your own kind are also equally self-aware. But Danny argued that gaining this useful ability would also result in understanding the *deaths* of others of your own kind—and, *consequently*, realizing one's own individual mortality. And he suggested that this all-encompassing, persistent, terror-filled realization would cause an individual who first made that critical step to lose out in the struggle to secure a mate and pass his or her genes to the next generation—in other words, such an individual would reach an evolutionary dead end. Danny suggested that we humans were the only species to finally get past this long-standing barrier. And he posited that we did this by *simultaneously* evolving mechanisms to deny our mortality.

I suspect most readers will have the same initial reaction I had—that this seems much more convoluted and complicated than simply saying that we humans evolved special mental abilities over time. But I realized that Danny was describing an apparently novel theory based on a counterintuitive line of logic, which seemed relevant to explaining both human origins and some unique features of the human mind. And my decade-long self-education about human origins had already prepared me to consider the larger implications of what he was saying. I began to think that such a rare and difficult transition might even possibly explain why all humans on the planet today seem to have emerged from a small group in Africa, completely replacing all other humanlike species that coexisted at the time. I was genuinely intrigued and excited and we spent the next hour in deep conversation, even after most others had left the lunch table. Despite our widely disparate backgrounds and education, Danny and I had one important thing in common: our shared understanding of a basic fact of evolutionary biology—that unless you are able to pass on your genes to the next generation by generating progeny, it does not matter how successful you were during your life. So any new genetically based abilities can become permanently established in a species only if they contribute to this "reproductive fitness."

The history of science is filled with ideas that initially went against conventional wisdom but were eventually proven to be true (such as Wegener's theory of continental drift and Copernicus's claim that the sun was the center of the solar system). So I was immediately attracted to Danny's contrarian idea of looking not for what *additional special features* of our brains made us human—but rather asking what might have *prevented* other animals and birds from becoming humanlike in their mental functions. In other words, had we crossed a very difficult evolutionary barrier on the way to becoming human? An analogy I later thought of was the process by which some ancient fishlike creatures moved from living in water to surviving on land. There were likely many attempts to make this transition, but evidence tells us that only a few such efforts actually succeeded. Apparently, several things had to happen at once, and in just the right order, to overcome this particular *physiological* evolutionary barrier. So why not also consider a *psychological* evolutionary barrier that blocks the path to humanlike awareness of reality? During our intense discussion I also asked Danny if religion could be the explanation for our success at overcoming that barrier, since all societies have religious belief systems and most religions provide explanations for what happens after death. He responded that while religion could have been a major factor that aided his proposed transition, it could not be the whole answer. After all, he said, do not most atheists live in constant terror of their own mortality? But he agreed that his theory could at least help explain the universality of religious belief systems in human societies. Most humans ask what lies beyond their death, and most religions provide an answer of some kind. There are also entire systems of philosophy that ask such existential questions, whether based on religious beliefs or not.

While there were obviously many details needed to support Danny's unusual line of thinking, I was impressed by the basic concept, and suggested that he should publish it. But I also realized that, like me, Danny had no prior formal education regarding human evolution, and in the academic world he would not be considered qualified to officially opine on the subject. His interest had simply grown out of his knowledge of evolutionary biology combined with the innate desire most humans have to understand our own origins.

On the face of it, this was not a momentous encounter—a conversation of less than two hours between two scientists from very different backgrounds, each with a nonexpert interest in explaining human origins. But over the months that followed I simply could not shake the basic idea Danny had proffered. The more I continued with my own quest to learn about human origins within the multidisciplinary CARTA group I had formed, the more this idea seemed to make sense and to gain in potential significance. After two years of obsessing about my discussion with Danny, I finally looked up his e-mail address and sent a lengthy message in which I outlined my understanding of his basic theory, updating and adding various embellishments of my own and suggesting again that he should publish his concept. I was deeply disappointed not to hear back from Danny, but thought that I might not have the right e-mail address. A few months later I decided to look up his phone number on the Internet, and was shocked to instead find his obituary. Danny Brower died suddenly and unexpectedly in October of 2007 from a rare kind of blood vessel disease called aortic dissection (possibly resulting from a defect in connective tissue molecules—the very things he had studied in flies). On a day he was due to present a departmental seminar, he woke up with severe symptoms for the very first time and went into surgery that evening. Tragically, he never regained consciousness, and was declared brain dead four days later.

Once I got over the shock of this unexpected and sad news, I scoured the published literature to see if Danny had ever written about his idea, but I found no evidence that he had. One day soon thereafter I saw a dedication to Danny in a research article on an unrelated topic by a well-known scientist named Sean Carroll.[15] I contacted Sean, who told me that

Danny had in fact talked to him and to a few other friends about some of his ideas. Sean had even read and commented on some writings Danny had begun on the subject—efforts cut short by Danny's untimely death. I was now even more convinced that the basic idea needed to be published. As it happened, I had previously gotten to know Philip Campbell, the editor in chief of the prestigious journal *Nature*, as he had once approached me to write an article about the ethics of doing research on great apes.[16] I contacted Campbell and explained the situation. He was interested and suggested that I write a formal "letter to the editor" on the topic.

Before writing the letter, I spent more time reading the literature and grew to appreciate the importance of a psychological concept called "theory of mind"—also variously called mind-reading, attribution of mental states, perspective taking, mindsight, and multilevel intentionality. These jargonistic terms refer to various aspects of the human ability to go beyond self-awareness of our own minds to the full comprehension that other humans are also self-aware and have independent minds of their own—and to thus put ourselves into their mental shoes. For example, the reason I could have a discussion with Danny was that we both knew that the other had a mind capable of independent thought and reasoning. And by now, you, the reader, may have started developing a theory of mind about both of us authors, including the one who is not even alive today.

I also consulted learned colleagues from CARTA in relevant disciplines to determine whether Danny's basic theory was truly original. It turned out that many other writers had already touched on the first half of the concept. Even ancient Indian Vedic texts had addressed the surprising fact that we humans deny the reality of our own mortality—easily—

though we know its certainty.[17] And in modern times, Ernest Becker's Pulitzer Prize–winning 1973 book *The Denial of Death* emphasized the point further, suggesting that many aspects of human behavior and culture can be explained by this denial mechanism.[18] But the second part of Danny's idea—that the realization of our own mortality might have been a barrier to the emergence of a humanlike mind until our species was finally able to deny that realization—was unique; I found nothing like it in anything I read. I wrote the letter to *Nature*, and it appeared in August of 2009.[19] The relevant sentences from the letter are reproduced below:

> Among key features of human uniqueness are full self-awareness and "theory of mind," which enables inter-subjectivity—an understanding of the intentionality of others. These attributes may have been positively selected because of their benefits to interpersonal communication, co-operative breeding, language and other critical human activities. However, the late Danny Brower, a geneticist from the University of Arizona, suggested to me that the real question is why they should have emerged in only one species, despite millions of years of opportunity. Here, I attempt to communicate Brower's concept. He explained that with full self-awareness and inter-subjectivity would also come awareness of death and mortality. Thus, far from being useful, the resulting overwhelming fear would be a dead-end evolutionary barrier, curbing activities and cognitive functions necessary for survival and reproductive fitness. Brower suggested that, although many species manifest features of self-awareness (including orangutans, chimpanzees, orcas, dolphins, elephants and perhaps magpies), the transition to a fully human-like phenotype was blocked for tens of mil-

lions of years of mammalian (and perhaps avian) evolution. In his view, the only way these properties could become positively selected was if they emerged simultaneously with neural mechanisms for denying mortality. Although aspects such as denial of death and awareness of mortality have been discussed as contributing to human culture and behaviour, to my knowledge Brower's concept of a long-standing evolutionary barrier had not previously been entertained. Brower's contrarian view could help modify and reinvigorate ongoing debates about the origins of human uniqueness and inter-subjectivity. It could also steer discussions of other uniquely human "universals," such as the ability to hold false beliefs, existential angst, theories of after-life, religiosity, severity of grieving, importance of death rituals, risk-taking behaviour, panic attacks, suicide and martyrdom. If this logic is correct, many warm-blooded species may have previously achieved complete self-awareness and inter-subjectivity, but then failed to survive because of the extremely negative immediate consequences. Perhaps we should be looking for the mechanisms (or loss of mechanisms) that allow us to delude ourselves and others about reality, even while realizing that both we and others are capable of such delusions and false beliefs.

Soon after the letter's publication, I heard from Sheldon Solomon, a member of a well-regarded group of psychologists influenced by the ideas of Ernest Becker and best known for their "terror management theory."[20] Their concept is supported by various types of experimental evidence and indicates that we humans have a variety of "worldview and self-esteem mechanisms" to deal with the terror of knowing we are going to die. In his letter to me, Solomon wrote: "We agree with your argument that the benefits of consciousness and self-awareness could only be reaped if they were accompanied by simultaneous mechanisms to deny death."

Thinking I had done my duty by getting Danny's ideas to the attention of others who could pursue them, I turned my focus to aspects of anthropogeny that were more directly related to my own expertise in glycosciences and ape-human differences in biology. But then I received a very unexpected e-mail from Danny's widow, Sharon Brower, who had been alerted to the *Nature* letter by one of Danny's friends. Sharon thanked me for bringing her late husband's unpublished idea into print, and told me that Danny had been spending all his spare time writing a book on the topic. Apparently, he had just completed his second draft before his sudden death. Ironically, while Danny had sometimes discussed plans for the distant future with Sharon, he also knew there was a possibility that he would die young: His own father had passed away suddenly at the age of fifty-six, of a heart attack. Danny's thinking about his theory may have even caused him to be more aware of his own mortality. According to Sharon, this scared Danny a bit; he had always held up fifty-six as the age to surpass. Sadly, he died just a month short of that milestone birthday.

When Sharon sent me her late husband's draft manuscript, I found that Danny not only had the core of an idea to explain the evolution of the human mind but that he also went on to present some important practical messages for humanity arising from his logic. He wrote that the human penchant to deny our mortality is but one manifestation of our overall ability to deny many other things—a propensity that has many ramifications, positive and negative. The manuscript was thoughtful and erudite, but it was clearly an incomplete effort that needed much additional research, expansion, and polishing, which Danny had been unable to do. With encouragement from Sharon, I therefore decided to continue the project by combining Danny's original writing with my own, adding thoughts, ideas, and embellishments along the way. In some places, I needed to simply correct or update issues that Danny did not have the time to finish researching. In other areas I added my own personal opinions and additions, with input from experts I consulted.[21]

In my first attempt, I simply could not bring myself to change any of Danny's original prose. Rather than alter his wording, I annotated his manuscript with extensive footnotes. While this helped me think through the whole concept, the product was not viable as a readable book. But some thoughtful readers advised me to follow Danny's intent—to write a book for a general audience, not a densely annotated scientific tome. And so (with Sharon's agreement) I decided to blend Danny's original text with my own additions, generating a text written mostly in one voice (throughout the rest of the book I will refer to myself in the first person where appropriate—in other instances, I will indicate when something refers specifically to Danny).

While I was working on the manuscript, my *Nature* letter was mentioned in a *Time* magazine cover article about the science of optimism by University College London neuroscientist Tali Sharot. This prelude to her book *The Optimism Bias* discusses the established fact that most humans maintain an irrationally positive outlook on life and asks how such optimism can be explained.[22] A stimulating e-mail discussion with Tali followed, which made me even more confident in thinking that Danny had been on to something important. After all, I thought, what is optimism but one form of denying reality? Meanwhile, a flood of other very relevant books kept appearing to complement the ones I already knew about.[23] While none of these books independently espoused the specific theory presented here, many seemed full of ideas and information directly or indirectly supportive of it. And my fifteen-year experience developing CARTA had given me the background needed to understand most of what these authors were saying.[24] It seemed as if fate had chosen me as the right person to complete Danny's work. Now I felt even more compelled to do so.

Neither Danny nor I began this quest with an advanced education in anthropogeny. We were both drawn to the subject by our curiosity and passion to explore the profound, universal human question of where we humans came from and how we got here. Readers who are not scientists also need not feel intimidated by the topic; any jargon that comes up is explained as we go along. Nor should you be put off by the apparent complexity of the matter—it befits exploration of one of the deepest and most fascinating universal human questions.

————

This book is about a proposed critical transition in the emergence of the human mind, a process that led to our becoming the dominant species on the planet. It also helps explain the mystery of why other highly intelligent mammals and birds have not developed humanlike mental abilities despite millions of years of opportunity to do so. The logic begins with the realization that even an animal with complete self-awareness cannot truly understand death until it becomes fully aware that others of its kind are also self-aware individuals—in other words, until it becomes aware of the personhood of others. This higher level of awareness is called a "full theory of mind," or the ability to fully "attribute mental states" to others, and with it comes an awareness of the deaths of others and thus the realization of one's own mortality. As we shall see later, the only way for a species to get past this death-anxiety barrier is by *denial* of this reality.

It is important at this point to note that the term *denial* has many different meanings, depending on the context. Standard dictionaries cite many different definitions of the word.[25] There are also diverse colloquial uses, including "self-denial" or "in denial," as well as various meanings in Freudian psychology. I use the term here not to denote dictionary definitions, such as "disbelief in the existence or reality of a thing" or "a refusal to agree or comply with a statement" (or other variants thereof), but rather to denote the basic definition derived from psychology: *An unconscious defense mechanism used to reduce anxiety by denying thoughts, feelings, or facts that are consciously intolerable.* This is akin to ignoring or not paying attention to something that is important. But the term *ignoring* implies a deliberate process, which is not necessary for our theory. The term *inattention* might also be substituted. But without a qualifier indicating whether the inattention is deliberate or spontaneous, it is difficult to apply this term as well. All in all, there is no perfect term to describe the human propensity that we are going to discuss. The phrases *denial of reality* or *reality denial* come as close as any. So this book is about how humans finally gained a full theory of mind by simultaneously attaining the ability to deny aspects of reality.

But wait, you say: *I* don't have much of a problem denying the reality of the fact that I am going to die. I can simply use my native human intelligence to consider the facts and statistics and rationalize away any fears I may have. After all, I might very well live a long time, so why should I worry about it right now? The fallacy in this logic is that you are a modern-day human whose ancestors have *already* crossed the barrier, and you are thus *already* capable of denying the unpleasant reality of your mortality. So it seems quite trivial to you. But as we shall see later, it was quite a different matter for the first humans who initially understood their mortality.

It is important to note that denial of mortality is also part of a much broader concept about our human ability to deny many other aspects of reality, especially when such realities are not to our liking. For example, we smoke cigarettes, eat unhealthful foods, don't watch our weight, and don't exercise, despite our full awareness that these habits are a prescription for an early death. And we go further in our denial of reality, holding completely false beliefs about many things, even in the face of the cold, hard facts. These uniquely human features have many implications—some rather negative—for our nature and behavior, and for many societal and global problems we face today. We've also mentioned the implications for the universal presence of religion in human populations, and the fact that almost all religions provide humans with some explanation regarding what happens after death.

After further reading and thinking, I could expand Danny's idea to suggest that humans breached a long-standing *psychological evolutionary barrier* as a result of a one-time-only confluence of events. In the process of thinking about it, I also came to realize that this theory could help explain a fascinating mystery about the origin of our own species—the fact that all of us apparently arose from a small group somewhere in Africa not so long ago and then spread across the planet, practically replacing all the other sophisticated humanlike species that had been coexisting with us. Later in the book I will suggest that we "behaviorally modern humans" were the only ones who actually made this evolutionarily difficult transition, leaving everyone else behind in the dust, mentally.

Last but not least, we will discuss some broader repercussions of our ability to deny the reality of anything that is unpleasant. We will point out that this human ability extends all the way from denial of personal health issues to our denial of the "mortality" of our planet's biosphere and climate in its present form. Even those of us who realize that the earth's climate and environment are being degraded by human activities tend to deny the urgency of the problem in our daily actions. This vital topic was discussed in Danny's draft chapters, and was another important reason why I felt that this book should be published and read by as many people as possible.

The book also brings together two disparate strands of human thought—one that seeks answers to the evolutionary mystery of the origins of the human mind and another that examines denial of reality (including denial of our mortality), which emerges out of the disciplines of existential philosophy, theology, psychoanalysis, and psychology. Each of these lines of thinking and research is already well developed in many existing tomes. But I believe this book brings them together for the first time, in an unexpected fusion. So please join Danny and me as we explore one of the greatest unsolved mysteries regarding our own species. Along the way we will encounter such diverse topics as Neandertal burials, irrational human optimism, Gandhi's principle of nonviolence, autism spectrum disorders, the relationship of dogs to humans, why elephants are killing cows and rhinos, the New Madrid Fault Line, skydiving accidents, Wallace's conundrum, and many other phenomena. But be forewarned that, like much of science, we will give you much to think about yet will not conclusively prove the theory we espouse. Rather, we will present "one long argument"[26] using facts and concepts in favor of our theory. But, in the tradition of science, we will also search assiduously for any "ugly facts" that might "slay our beautiful hypothesis."[27]

It is not often that one's life is so dramatically affected by a single chance meeting with one individual, especially a meeting that lasted less than two hours. This book is presented in homage to Danny's original thinking, as it has helped me to illuminate not only the matter of human origins and the human condition but also has major implications for the future of our species and of our planet. It is a wake-up call to all of humanity about our remarkable ability for reality denial—which can either continue to lead to our downfall or turn into one of our greatest assets.

CHAPTER 1

Where Did We Come From, and How Did We Get Here?

Facts are stubborn things; and whatever may be our wishes, our inclinations, or the dictates of our passions, they cannot alter the state of facts and evidence.
—John Adams, speech in defense of the soldiers at the Boston Massacre trials

Science tells us that we are creatures of accident clinging to a ball of mud hurtling aimlessly through space. This is not a notion to warm hearts or rouse multitudes.
—Paul Ehrlich, in *Human Natures*

At the public symposia organized by CARTA,[1] we usually begin by rhetorically asking humanity's oldest questions—Who are we? What are we doing here? Where did we come from? How did we get here? Where are we going? We then emphasize that while some of these questions fall under the domains of philosophy and religion, two are open to rigorous scientific inquiry—where we came from and how we got here. And these questions, of course, comprise the primary focus of CARTA, namely, anthropogeny. Before diving into a discussion of our theory about the origins of the human mind, though, we must ensure that the reader is up to speed on what is known to date about where our species came from. And to consider how we got here, we must also review the fundamental principles of biological evolution— because we humans are a product of this remarkable process. We will then close this chapter by discussing a little about what is known about the human mind in order to put our theory into perspective. But we will be very brief on all three counts, as many other books have addressed such matters in excellent detail.

Let's begin with a general summary of knowledge about human origins as of the end of 2012.[2] As you probably know, we humans are warm-blooded animals called mammals, whose ancestors existed before the dinosaurs disappeared around sixty-five million years ago, after a giant comet hit the earth. Among the mammals that survived and evolved, we emerged as primates, part of a subgroup called Old World primates, which arose in Africa and/or Asia around thirty to forty million years ago. Humans belong to a still smaller sub-set called apes, which are characterized by the lack of a tail. Among the apes, we are derived from a subgroup tradition-ally called the great apes, of which the other currently living species are chimpanzees, bonobos, gorillas, and orangutans.

The term *great ape* has now fallen out of favor, though, because it turns out that we are closer genetically to chim-panzees and bonobos (so-called pygmy chimpanzees) than they are to gorillas and orangutans. Indeed, Jared Diamond was spot-on in saying that, from a strictly genetic perspec-tive, we are nothing more than "a third chimpanzee."[3] The latest taxonomic classification, therefore, lumps all great apes and humans together as hominids, in recognition of our ge-netic similarities. The fossil species that appeared *after* our common ancestor with chimpanzees appeared are thus now called hominins, rather than hominids. To place all this in perspective, consider the fact mentioned earlier, that we are genetically closer to chimpanzees than mice and rats are to each other!

The earliest hominin fossils have all been found in Africa, dating back as far as around five to seven million years ago, presumably close to the time of our common ancestor with chimpanzees.[4] The location and physical appearance of this common ancestor remains unknown, but fossils suggest that it was a bit more chimpanzee-like than humanlike, especially with regard to its smaller skull case. Apart from the emergence of an upright posture and bipedalism (walking on two legs), these species do not appear to have undergone very major humanlike changes for another three or four million years. Rather, they seem to have gone through various specializations, and a bushy, branching tree of various species apparently coexisted at the time. The best known of these species is Ardi, a more than four-million-year-old skeleton classified as *Ardipithecus ramidus*, which was described by Tim White and his colleagues,[5] and Lucy, a more than three-million-year-old skeleton classified as *Australopithecus afarensis*, which was discovered by Don Johanson and his colleagues.[6] Starting about two million years ago, one sees fossil evidence of species discovered by Louis and Mary Leakey and others classified under the genus *Homo*, in which there is some increase in brain size and the beginnings of stone tool use. A species called *Homo ergaster* then emerges in Africa, exemplified by the skeleton called Turkana Boy, discovered by Richard Leakey and colleagues,[7] which is bigger in body size and appears committed to striding, bipedal walking, and, likely, long-distance running.[8]

Along with this change came a gradual increase in brain size, an increased sophistication in the making and use of stone tools, and evidence of more meat eating. Very similar creatures called *Homo erectus* are then found spreading throughout the Old World, migrating as far as present-day Indonesia, China, and Europe.

Over the next two million years or so *Homo ergaster/erectus* seems to have undergone only minor changes, other than further increases in brain size and limited improvements in stone tools. Beginning about a million years ago, we see evidence of continuing brain size increase in *Homo*, with the modern size being achieved around three or four hundred thousand years ago. But at that point there is still no archaeological evidence of what looks like so-called modern (present-day) human behavior, such as burials with funerary artifacts, representative drawings, ornaments, trade, and so on. Thus it seems that the final increase in brain size was not sufficient to allow the emergence of us "modern humans." One of the best-documented species from this period are Neandertals,[9] who were found throughout what is now Europe, the Middle East, and western Asia, often living in extreme climates associated with ice ages. These closest extinct evolutionary relatives were short, squat powerhouses with robust skeletal anatomy; their brains were larger than ours are even now, they used more specialized and complex stone tools than *Homo erectus*, and they were able to control fire. However, the first two hundred thousand years or so of their existence were not characterized by the kinds of artifacts associated with modern human behaviors, such as bead necklaces, symbolic art, and the like. The recent discovery of Denisovans (defined by a few bones found in a Siberian cave and the genomic DNA from these samples) has uncovered another branch of ancient humans, which likely shared common ancestors with humans and Neandertals a few hundred thousand years ago.

Skeletons very similar to those of present-day humans can be found in Africa, beginning about two hundred thousand years ago. However, these "anatomically modern humans"[10] also did not leave behind many artifacts suggestive of modern human behavior (although there is some indirect evidence of symbolic thinking, such as the production of ochre colors that were presumably used for body decoration). We "behaviorally modern humans," then emerged somewhere in Africa around one hundred thousand years ago[11] (and DNA analysis of current-day humans says that we all came from a relatively small effective population size, which numbered only around five or ten thousand individuals or so). The first evidence for truly modern human behavior—evidence such as complex symmetric scratch marks on objects, bead necklaces, and burials with funerary goods—dates to about this time. Shortly thereafter some of us began to leave Africa (some may have also migrated back). These migrations appear to have first taken humans into what is now Palestine and Israel, and then along the coastlines of the Middle East, India, and Indonesia—eventually leading them into Australia (one of the first crossings of deep water in which the horizon would have shown no evidence of land on the other side). Around this time our ancestors also began appearing in what is now China and southern Europe. The later European Cro-Magnons are incorrectly claimed to be the first human group to demonstrate modern human behavior. What was once thought of as a cultural "great leap forward" by these early Europeans (who made bead necklaces, spear-throwing implements, cave art, and performed ceremonial burials) is now felt to be a result of improved preservation and more sustained research in this particular region.

As we migrated across the world, various human subgroups evolved changes in superficial aspects of their appearance, such as skin and eye color. There was also further local evolution, such as the ability of some populations to digest milk or others to become resistant to some diseases. But today we move children from one part of the world to another and see them perform all human mental activities in a very similar fashion (the United States has inadvertently done this experiment on a broad scale via its immigration policies). So we humans are all far, far more similar than we are different from each other.[12] And all modern human mental abilities must have been fully established in our ancestors before their initial migration out of Africa. Interestingly, after coexisting for another fifty thousand years or so, all other humanlike species disappeared, leaving only us "behaviorally modern humans" behind.

What about other defining human characteristics? Apart from features such as bipedalism, increased brain size, and development of the human type of opposing digits that are detectable in the fossil record, we know little about when and where other human-specific features, such as hairlessness, emerged. Likewise, details regarding the first development of perishable artifacts, such as clothing, ropes, nets, etc., remain largely unknown (although clues abound). Last but not least, it is unclear when and where the abilities for human language, full theory of mind, and other advanced cognitive capabilities emerged, except that they must have predated our initial diaspora from Africa. This book considers *how* these unusual and powerful abilities may have emerged in

Africa by first asking *why* they did not also emerge in any other species, despite tens of millions of years of opportunities to do so.

Having defined *who* we are, let's move on to the question of *how* we got here. Humans are the product of biological evolution—so a basic understanding of this process and how it generates species is important if one is to appreciate the rest of this book. But some people still resist the notion that they shared a common ancestor with a chimpanzee. In other words, they do not accept what scientific exploration has revealed. Science uncovers the reality of things around us and how they work. New discoveries often just raise even more questions. But scientific conclusions sometimes rise to the level of indisputable fact. For example, we once thought the earth was flat, and that it was the center of our universe. But we now accept that the earth is round, and that it orbits a minor star called the sun. In like manner, biological evolution by natural selection is now established *as an incontrovertible fact.* The only reason we refer to evolution as a "theory" is that biologists are perhaps more reticent than physicists, who might have declared a "law" of evolution long ago. Experts do still argue about details, but evidence for the fundamental tenets of the theory of evolution by natural selection is so overwhelming that no reputable biologist disputes it. It also underlies much modern biological and biomedical research. And by the end of this book, we hope to convince you that understanding evolution and its repercussions is also critical for our survival as a species.

Charles Darwin and Alfred Russel Wallace discovered the principle of evolution via natural selection independently, but initially published their findings together. How this coincidence came about is a fascinating story itself, but we will not go into that matter here.[13] Combining many insights,[14] these two visionaries realized that

1. resources are limited;
2. members of a population of organisms are different from each other (genetically heterogeneous);
3. these differences make some individuals better able to compete for resources;
4. genetic traits are passed on to one's progeny;
5. heritable traits of better competitors will become more common; and
6. organisms will therefore change over time, either because different conditions select for particular traits or because genetic changes keep occurring, generating better "adapted" organisms.

These are the basic components of biological evolution by natural selection. The underlying concept is so simple that Darwin's friend Thomas Huxley later remarked, "How stupid of me not to have thought of that!"

But contrary to the popular misconception, Darwin did not use the phrase "survival of the fittest" until later editions of his famous book *On the Origin of Species*. This catchy expression was actually coined by the economist and philosopher Herbert Spencer in his 1864 book, *The Principles of Biology*, because he mistakenly saw parallels between his economic theories and Darwin's biological theory of natural selection. And we now know from the insights of Motoo Kimura and others that natural selection is happening against a backdrop of neutral evolution (random changes that are not detrimental to survival and reproduction and that are able to persist for this reason). In addition, depending on the circumstances, there can be "survival of the luckiest."

So if a species will thus change over time, how does an entirely new species emerge? A precise definition of "species" is actually tricky, as evolution has no "plan" to generate species. Humans have traditionally defined a species as a population in which two individuals are able to mate and produce fertile offspring.[15] If two populations of a species become isolated from each other by a geographic barrier, such as an ocean or mountain range, they can evolve independently, changing genetically in ways that make them different enough to become separate species.[16] But species can also split if genes cannot pass between groups frequently enough, even though they occupy the same territory.[17] There are many ways in which this kind of "sympatric speciation" can occur, and many opinions regarding their relative importance. For just one perspective, see Etienne Joly's essay in which he suggests that new species arise not by populations splitting into separate branches but by small inbreeding groups "budding" from an ancestral stock.[18] A new species can also emerge because of a "bottleneck," in which environmental factors cause a constriction of a population into a small group—as may have happened at the origins of our own species (more on this later). In relation to earlier stages in human evolution, Pascal Gagneux at UC San Diego has even shown that the evolutionary loss of one kind of sialic acid (a sugar on the surface of cells) in our human ancestors about two million years ago induced females to make antibodies against it, which then attacked the sperm of males who still had it, causing partial infertility.[19] This particular immune obstacle to reproduction may have been sufficient to induce the speciation of the genus *Homo*, the lineage that eventually gave rise to us, two million years later.

We now know in great detail what genes are made of, how genetic information is copied when each cell divides, and how these instructions are translated into specific proteins, whose activities (interacting with other molecules and with the environment) are eventually manifested as traits. We can also measure the rate at which errors in DNA copying lead to mutations (errors) in eggs and sperm, generating variations that allow different members of a population to compete better, or less well, for resources. There is plenty of genetic variability that evolutionary selection can "choose" from. Indeed, based on our current molecular knowledge of life, *evolution by natural selection is unavoidable*, and is a *defining attribute of all life forms*.[20] The most obvious evidence is the fossil record. But this record is incomplete, and a lack of so-called missing links is sometimes cited by evolution doubters as evidence for permanence of species. However, as the number of known fossils increases, evolutionary intermediates are emerging.[21] Another point is that evolution typically takes an existing structure and modifies it for new uses, as opposed to generating new things from scratch.[22] We can also see evolution happening naturally in real time.[23] And there are examples of selection important for health, such as our ongoing battle with infectious agents,[24] bacteria's resistance to antibiotics,[25] the rapid emergence of new influenza strains,[26] and the ever-changing HIV virus, which causes AIDS.[27] With such short-generation-time microorganisms, selection can occur with astonishing speed. But with longer-lived organisms it is more difficult to see selection leading to speciation (derivation of two species from one) over short time frames. It is easier to identify organisms in the various phases of speciation.

In the final analysis, "species" is just a concept invented by humans in an attempt to come to grips with the vastness of biological diversity generated by the evolutionary process. Biologists who specialize in classification of organisms (systematics) thus also speak of "subspecies" based on traits such as color, size, etc.[28] This ambiguity in designating "species" itself demonstrates that life is evolving and changing into new forms even as we speak. Indeed, it appears that when our own species began to migrate from Africa tens of thousands of years ago we met and bred to a limited extent with our then closest living evolutionary cousins, the Neandertals and Denisovans, who stayed around until just a few tens of thousands of years ago. While Michael Crichton's *Jurassic Park* was just science fiction, Svante Pääbo and his colleagues have actually extracted DNA from the fossil remains of these extinct humanlike species. So we now know that most people from Europe and Asia have a little bit of Neandertal DNA in their genomes,[29] and some people from Southeast Asia carry bits from Denisovans.[30] But the extent of this interbreeding seems to have been limited. We will return to this later, as the theory in this book may provide the answer to what the limitation might have been.

What else do you need to know about biological evolution in order to understand this book? One important concept is that although natural selection acts on individuals, it eventually shapes populations of organisms, which are made up of a collection of traits encoded in genes.[31] Of course, the inheritance of most traits is not straightforward. For example, your height is determined by the actions of many, many genes and is also affected by your life circumstances. Especially with humans, it is difficult to attribute anything just to genes, as the environment plays a strong role in determining the final outcome. As the title of Matt Ridley's book put it, it is *Nature via Nurture*,[32] not nature *versus* nurture—it is past time to put an end to this false dichotomy.[33] And your genes are *not a blueprint* for what you are, they are just the entries in a *recipe book* that both defines and limits the options for what you can become. Of course, while your environment can change your own characteristics, it does not directly change the gene sequences that are passed on to your progeny. Your genes interact with your environment to make you what you are, but your basic genetic endowment does not change. Interaction with the environment simply affects the chances that your genes will make it into the next generation.[34]

Thousands of genes, trillions of cells, mutations, etc., etc.—you may be worrying that this is all going to be too complicated. Don't give up just now, because we are about to make it all much simpler. While natural selection works on all your genes and the way they interact with your environment, it is integrating all these individual inputs to produce *one unifying composite trait that matters most*, called fitness. Your fitness is defined simply as the likelihood that your genes will find their way into the next generation—how successfully you reproduce compared with others in the population. If the sum of all your traits makes it more likely that you will have more descendants than your neighbor, then your fitness is greater. Thus *relative reproductive fitness is the single overriding trait being gauged by natural selection*, and anything that affects your chances of mating successfully will be an important component of fitness.[35]

Because of this fact, evolution can even generate unusual features that at first glance appear to be pitted against fitness, such as traits selected by mating rituals, which are part of what is called sexual selection. It is hard to imagine that the exaggerated tail feathers of a male peacock give the bird any survival advantage. Indeed, dragging around such an outlandish appendage seems only to be a detriment to survival. However, the male that can best strut his stuff with his fantastic, quivering fan of tail feathers is going to be more successful at convincing Miss Peahen that he is the guy for her. And unless he is successful, his fitness takes a nosedive. This is an extreme example of what sexual selection can do, seeming to run almost counter to conventional natural selection by environmental factors. But both processes actually work together toward increasing reproductive fitness. The mating ritual provides many other examples of situations in which traits can present immediate conflicts between survival and fitness. Consider animals in which a dominant male has a harem of females. In such cases, a male is required to fight the reigning supremo for dominance and the privilege of spreading his genes among the ladies. Dominance contests are often ritualized to limit damage, but in many species the danger of serious harm or death is very real. Is it in the best interest of self-survival to take part in such a contest? Of course not—but if a male is unwilling to engage in risky combat, his fitness is close to zero.[36] More on this later, when we consider what might happen to these innate sexual drives when one member of a species begins to fully understand the minds of others.

In summary, natural selection ultimately acts to maximize reproductive fitness. Your genes must get into the next generation or they will disappear. For most animals, reproductive fitness and survival share largely overlapping goals during early growth and development—to reach maturity with the best equipment possible to ensure reproduction. However, once this stage is reached, the interests of fitness and survival become divergent, and evolution is going to favor traits that maximize your ability to leave successful progeny, even if those traits carry disadvantages to your own long-term self-preservation.[37]

Senator Daniel Patrick Moynihan once said: *"Everyone is entitled to his own opinion, but not his own facts."* We hope by this point that all readers understand and accept these basic facts about evolution. Regardless of any faith-based belief system, you should be willing to accept the reality of scientific facts as they are discovered and adjust your beliefs accordingly. Indeed, if you do believe in a creator who gave us humans the ability to discover facts about nature, why would you then disbelieve what this God-given ability has revealed? Furthermore, biology is chock-full of unintelligent "design"—just look at how badly the human body is designed.[38] So anyone who insists that God designed and created this very messy biology from scratch has a poor opinion of God's abilities.

To avoid the nonscientific arguments that continue to surround this sociopolitically contentious issue, I recently derived a conclusive Proof of Evolution by First Principles, which incorporates well-known facts and doesn't allude to specific theories or to any literature on evolution per se. A reader who is an evolution skeptic should read this complete proof in the endnote carefully.[39] This knowledge is essential to fully appreciate the rest of this book.

Finally, since this book is about a major psychological transition during human evolution, it is about the human brain. So we also need to review some very basic facts about this three-pound blob of jelly inside your skull, the ultimate source of your mind and your self. The human brain contains more nerve cells than there are stars in the Milky Way (more than one hundred billion), and each of these cells may have a thousand or more connections to other cells. Given this overwhelming complexity, we have a long, long way to go to really understand the human brain and how it generates the mind.[40] But there is a huge amount of information already known in the neurosciences, and it is impossible to cover even the basics here. Fortunately, the specifics of the brain mechanisms involved do not have to be fully understood in order to appreciate the theory presented in this book. Interested readers are referred to the many excellent books and articles about the human brain and mind, its complexities, quirks, and underlying mechanisms.[41]

Is who we become determined by our genes or by our upbringing? The short answer is both. As we said, DNA is not a blueprint, just a recipe book that limits your options. In keeping with this, study after study (many based on twins who happened to have been separated and raised in different environments) indicate that genetics plays a major if not predominant role in personality.[42] On the other hand, the environment has a major influence on how your mind develops after birth. Part of the problem with the false nature-nurture dichotomy is that we lose sight of the fact that personality is very different from behavior. Your personality is the foundation, the base, from which you build a pattern of behaviors—as Jerome Kagan emphasizes, how you turn out is much affected by the temperament you are born with.[43] Many studies that demonstrate the predominance of genetics typically measure aspects of our temperament, personalities, and raw intelligence, traits that seem to be mostly under the influence of our genes. But our upbringing (nurture) is going to have a huge impact on how we channel our personality traits and intelligence into behavior and intellectual achievements and on how we develop our emotional intelligence.[44] Two humans with identical personality traits could grow up to be very different people if exposed to differing environments. I once suggested a fictitious example of genetically identical Japanese twins who go their separate ways, one eventually becoming a sumo wrestler and the other an ascetic Buddhist monk.[45] Despite their identical genetic background, some hypothetical alien anthropologist observing them later in life might well conclude that they actually belong to different species! But on careful examination they will likely have many similarities in likes, dislikes, preferences, and personality traits.

There is another reason people have difficulty believing our behaviors are influenced by natural selection—we look around and see others doing so many stupid or even self-destructive things that it's hard to believe these could be favored by selection (amusing but true instances are featured in the tongue-in-cheek Darwin Awards, given to "salute the improvement of the human genome by honoring those who accidentally remove themselves from it").[46] Yes, it is true that we don't often do what is best for us, and we will discuss some reasons for that later, in relation to the human penchant for denying reality. But who we are is also not a result of selection in the world we currently inhabit. Human behavioral evolution was shaped by selection in environments we lived in prior to the advent of civilization, which likely included small groups with relatively simple social structures. Humans maintained this relatively simple existence for tens of thousands of years. Until the advent of agriculture, less than ten thousand years ago, the large cities that threw together multitudes did not exist. Prehistoric humans were able to pass down the wisdom of learned experiences from generation to generation by oral traditions, but once towns and cities were created and writing was invented, the pace of human social and cultural evolution picked up, aided by written laws and codes of conduct.[47] Colin Renfrew has called this *"the sapient paradox...the gap between the appearance of modern humans and the range of new behaviors associated with the agricultural revolution."*[48] These human successes can be largely attributed to our capacity for "cumulative culture," with its ever-accelerating accumulation of knowledge and technology over time.[49] A recent study of problem solving compared monkeys, chimpanzees, and children, showing that the children's much greater success was attributable to factors such as teaching through verbal instruction, imitation, and the tendency to cooperate.[50]

In the end, what differentiates us most from other animals seems to be several such unusual mental abilities, which seem to have appeared not so long ago. As Ian Tattersall puts it: "The changeover of *Homo sapiens* from a nonsymbolic, nonlinguistic species to a symbolic, linguistic one is the most mind-boggling cognitive transformation that has ever happened to any organism."[51] William Calvin calls this the big bang of the human mind.[52] Is this really a qualitative difference, or just a difference of scale so great that it looks like a difference in type? In fact, there is as yet no evidence that our brains themselves are fundamentally different from those of related mammals. While relevant differences exist, most of the anatomical structures of our brains have corresponding components in other mammals, even rodents.

Ah, but we are rational beings, unlike those others, you say. Well, just ask any practicing psychologist or psychiatrist how rational we are. We are emotional beings, and, as we will discuss later, the ability of our emotions to short-circuit rationality was probably an essential part of our evolutionary development. Other apparent nonselective behaviors result from the fact that situations arise in which a behavior that is usually good for fitness can become bad. Overall, selection will fix those traits that optimize fitness of the greatest number of individuals in the population. But under unusual circumstances, these generally positive behaviors can be counterselective. For example, most humans (but not most other animals) will love and nurture an offspring that is physically or mentally disabled and has little chance of contributing to the propagation of the genetic heritage of the parents. This expenditure of valuable resources is not in the best interest of the parents' long-term "fitness." However, evolution has selected strongly for unconditional love of offspring by humans (particularly mothers). Let's face it: Without such empathy, it would be hard to put up with the constant attention that youngsters require. We don't ask if there is going to be a payoff in fitness—we are evolved to do it. In the specific case of a badly disabled child, this trait is counterselective. But this selective disadvantage is outweighed in a few cases by the advantage of unconditional parental love in the population as a whole. And while the child may not contribute genes to the next generation, she or he may contribute in other domains, such as creativity and culture.

So the human mind has been shaped by selective forces, which have attempted to maximize fitness in the population as a whole. Aberrations are often easily seen to be just that— behaviors that are not good in a specific circumstance even though they are beneficial on the whole. And behaviors that seem to be less than optimal can often be explained by the differences between our current environment and the environment that we experienced when our brains were being molded, thousands of years ago. The important thing is that apparent exceptions in isolated cases must be viewed as part of the larger picture.

But one thing that does seem to separate us from other animals is that we are capable of going beyond self-awareness of our own personhood to having a full theory of mind (i.e., full *awareness of the self-awareness and personhood of others*). This apparently unique human ability is addressed at length in chapter 4. For now, let us just say that we don't think human brains are fundamentally different from those of other species, just differently developed. But then why do we appear so mentally different from the many other smart creatures on the planet? How and when did we make that critical transition?

CHAPTER 2

Becoming Smarter Shouldn't Be Hard

> We are a fluke of nature, a quirk of evolution, a glorious contingency.
>
> —Michael Shermer, in *The Believing Brain*

This book proceeds on the assumption that something rather unusual happened during the evolution of the human mind, and it suggests a novel explanation. To be clear at the outset, we are not saying that humans are "special" compared to other animals. That is a matter of personal opinion. In fact, many aspects of our human senses and physical abilities seem inferior to those of many other animals, as we will discuss later. But our attitude toward the question of human uniqueness has undergone a major change over the last century.[1] We have gone from the old notion that humans are at the peak of God's creation and vastly different from all other animals (i.e., that we are at the pinnacle of the *scala naturae*, or great chain of being)—all the way to the currently popular notion that we are "just another ape" and that there is nothing at all unusual about us. The reasons for this "humanization" of other animals are many, including a perfectly legitimate desire to treat them better than we currently do.

Popular books by Jane Goodall,[2] Frans de Waal,[3] Bill McGrew,[4] Tetsuro Matsuzawa,[5] and others advance the view that humans are remarkably similar to other primates, a position that is indeed backed by much scientific evidence. However, we feel that this view has been taken to extremes by some others, who would suggest there is nothing at all unusual about the human mind. The pendulum has swung too far in that direction. If the human mind was not unusual, we would not be writing this book and you would not be reading it, thinking about the matter, and possibly discussing its content with others.

While many current authors emphasize the extreme similarity of humans to our closest evolutionary cousins, a few have taken an opposing view. Danny Povinelli originally decided to become an anthropologist and study chimpanzees because he was fascinated as a child by popular media reports suggesting that they were mentally very similar to humans. His surprise and disappointment when his own studies showed that this is not the case make his current views[6] all the more compelling. Michael Tomasello comes to similar though not identical conclusions about the cognitive abilities of chimpanzees.[7] Meanwhile, an older generation of experts, including David Premack[8] and Herb Terrace,[9] are saying much the same thing as they look back at their own experiences of studying chimpanzees. And, like the little boy who pointed out that the emperor had no clothes, science writer Jon Cohen[10] and film producer Jeremy Taylor[11] (who have no vested interest or academic background in the topic) are now questioning this "chimpanization" of humans. As nonexperts on chimpanzees, both these authors spent much time researching the matter and came to the same general conclusion—that there is something extraordinary about the human mind in comparison to our closest evolutionary cousins. Raymond Tallis comes to similar conclusions in *Aping Mankind*.[12] Indeed, you only have to ponder for a moment what you're doing right now and what you're about to read to realize that there is something that needs explanation. Now, if you're still among those who insist that there is nothing at all unusual about the human mind compared to those of other animals, then this book will not be of interest to you. On the other hand, if you believe that the human mind can do many things other animals are incapable of, there is a mystery to be explored. Why and how did this come about? Let us now approach this issue from a new perspective.

At first glance, it would seem that becoming more intelligent should be a good thing during evolution, and that it should not be too hard. As we already discussed, evolution often works by positive selection, which means that a particular trait (feature) of a species is favored because it confers some fitness advantage in some context. And contrary to the popular misconception, fitness does not primarily denote physical or mental fitness. Rather, what really matters is one's ability to produce more offspring than others in the same population—that is, reproductive fitness in passing on your positively selected traits and genes to your progeny. Perhaps the more accurate phrase is "survival of the fittest reproducers."

So what factors might have provided a positive selection for increased intelligence? This may seem simple, but it is in fact not obvious that having humanlike intelligence is such an excellent strategy. Yes, we may appear to be the dominant species on the earth at this time, but many other species have adapted quite well without being smart enough to write *War and Peace*. Cockroaches have been around for many, many more millions of years than we have, and they are likely to be scurrying about long after we are gone. If one does a quick mental survey of all animals, there actually does not appear to be much of a correlation between overall intelligence and long-term success as a species (in terms of numbers and longevity). The reality is that while intelligence comes with many benefits, it also has some potential downsides.

So greater intelligence must have arisen in humans because it allowed us to better occupy a particular niche in our environment. What is a niche? In general, a species' niche is a description of the place that it occupies in an ecosystem (the sum total of interactions between a community of living organisms in a particular area and their nonliving environment). The niche includes all the things that an animal eats, where it lives, when it is active, etc. If a critter is occupying a niche that mostly entails burrowing through soil and eating worms that are detected by smell and touch, then it is not obvious that being smart is going to be particularly useful. This is especially true if being smart requires a large head (which would require excavating bigger tunnels for a burrowing animal) and a very active metabolism (which would not be easy to maintain unless there are a lot of worms to eat). So it is actually not surprising that most animals are not very smart by human standards. Indeed, many very successful species appear rather dumb from the human perspective, even by comparison to most of our four-legged friends.

Although being very smart is not a common path to evolutionary success, it can be one of a number of useful strategies, depending on the niche occupied by a species and other biological factors. Certainly being smarter can potentially help an animal be more successful at extracting resources from its environment or avoiding predators. But a smarter giraffe is not going to be able to climb up a tree to reach even higher leaves. Being smarter will not help a desert toad survive the long intervals between rainy periods. Only a certain amount of neural wiring is necessary in order to make an animal's behavior quite efficient for the life that it is designed to live in the niche that it occupies (note that the word *designed* here does not mean there was a "designer" involved—processes of evolution over long periods of time simply generate structures and functions so remarkably efficient that they give the appearance of being designed). Nature has not found it necessary to evolve brains capable of advanced thought processes, such as meditative reflection, in order to maximize behaviors that can find food, fend off competition, avoid being eaten, and reproduce successfully (sometimes facetiously called the famous four *f*s: feeding, fighting, fleeing, and fornication).

So what conditions would be likely to select for increased intelligence? Animals are not just being selected for survival based on their ability to deal with other species and environmental factors. Many are also often in competition with other members of their own species. As Darwin realized, the "struggle for existence" not only involves competition with other species but also reflects on the abilities of a given individual in a species to survive and reproduce compared to other individuals in the same population. Some experts who ponder the merits of brains for improving fitness think that for the very smartest animals, the primary selective advantage of increased intelligence comes from a better ability to compete against other individuals within the species, which may also involve the ability to interact most successfully in social situations.[13] In various versions of this social selection scenario, natural selection doesn't act to defend a niche from other animals but rather renders one individual more successful than others who occupy the same niche. Such "cognitive competition" within a species is one factor underlying the theory presented in this book. Of course, in humans, competition between groups can be a major added factor.

When dealing with competition within a species, powerful new selective factors enter the equation. Wolves may compete with other animals (e.g., cougars) for food, shelter, and other resources. However, they also compete with other wolves for mates (wolves and cougars, of course, do not mate with each other). And mating success also weighs very heavily on reproductive fitness; mate more, pass more of one's genes on to the next generation. Selection for mates throws new complexities into any consideration of selective advantage. As we have already mentioned, such sexual selection is a well-recognized force in evolution. For multiple reasons that we will not go into here (see Matt Ridley's *The Red Queen*[14] for an excellent discussion), almost all multicellular animals engage in the less efficient and more risky process of sexual reproduction rather than the simpler process of asexual reproduction (e.g., just splitting in two, as a single-celled amoeba does). Sexual selection in the form of mate choice can result in the evolution of remarkable phenomena that go far beyond what should be needed for survival. As we said earlier, sexual selection can sometimes run apparently counter to natural selection, actually reducing the ability of an individual to survive. The flamboyant peacock's tail evolved to attract females, not to fly—and in fact hobbles the bird as it tries to get away from predators. The elaborate and aesthetically beautiful (to us) "bowers" constructed by New Guinean and Australian male bowerbirds to impress females is another example of the expenditure of energy that individuals will endure in the effort to pass on their genes (there are eighteen species of bowerbirds, and the bowers tend to be species-specific in design).[15] At the level of male-male physical competition for mating partners, the ungainly and massive antlers of a moose come to mind. Of course, for those organisms in which intelligence is prominent, there is also a potential for sexual selection to be based on a mating preference for better mental abilities. In other words, mental prowess could be the invisible equivalent of a peacock's tail, but would require corresponding mental abilities to be correctly gauged by the opposite sex.

We can see how sexual selection produces surprising morphological or behavioral effects in some animals that would make little sense in the context of maximizing a competitive advantage against other species. Sexual selection can waste considerable resources, or it can saddle an adult with remarkable physical handicaps, in the effort to increase the chances of wooing a mate. This should come as no surprise, considering the importance of mating in the continuation of one's genetic legacy. Mate selection and access to mates also complicate other, more straightforward, factors that affect fitness. Food and shelter, for example, will help an individual survive, but accumulation of these resources can also be important in making one attractive to prospective mates. Particularly in a species in which both parents can contribute to raising offspring, females may be inclined to use resource accumulation to indicate which males will be the best fathers and providers. Thus one's status in the social pecking order can feed into the fitness equation in many interdependent ways. Of course, brute force, as well as access to mates, can instead determine the social pecking order.

Which types of animals would you expect to be most strongly selected for intelligence? That is, what type of niche would most reward an animal that is a little smarter than its brethren? We can all agree that life becomes more complicated as our social structures become more complex. Animals that live in groups often form dominance hierarchies, hunt in collaborative packs, and so on. All this requires relatively sophisticated mental processing, and it is easy to see that being more intelligent than the neighboring pack could be advantageous. The ability to communicate with one another with precision is also important in groups, and this helps to drive intelligence. Then consider the competition between individuals within a pack of mammals. If one male is better at making decisions, finding food, and choosing the best shelter, that individual has a higher probability of becoming a leader. In a social group of animals, one of the typical perks of higher status is more mating opportunities. Females are genetically primed by evolution to unconsciously "choose" mates offering the best genes for their progeny, and they often use status in the group as an important indicator of this genetic prowess. One might also expect that being smarter should have a greater benefit if an animal is relatively long-lived, giving it more time to maximize the use of such abilities.

Although there are exceptions to every rule in biology,[16] it is indeed the case that large-brained socially complex animals (such as humans, chimpanzees, dolphins, and elephants) tend to be relatively long-lived. And this package of characteristics often includes a slower rate of development after birth as well as the opportunity to learn more during the time before reaching adulthood. But learning does little good if you don't live long enough to use what you have learned and then pass your genes on to the next generation. There are some relatively trivial constraints as well. A certain brain size is required—it's hard to imagine that an ant's head could easily accommodate the number of cells required to learn quantum physics. Brains also use a lot of energy, and the incredibly complex networks of interacting neurons will work best if conditions (such as temperature) are kept relatively constant. Thus we would expect evolution for intelligence to be best developed in animals that maintain a relatively constant body temperature.

The animals we typically consider to be very intelligent happen to meet the above criteria. Elephants, cetaceans (whales and dolphins), and other primates (monkeys and apes) are all pretty smart as animals go, and all establish social groups, enjoy multiyear life spans, keep a constant body temperature, and have sufficiently large heads. This concept of a large head (big brain) is, of course, a relative one. Elephants and whales have much larger brains than those of humans, but when brain size is considered relative to body size, humans do come out a bit ahead of the pack.[17] On the other hand, although having a larger brain was apparently a necessary step toward developing humanlike mental abilities, it was evidently not a sufficient one. Indeed, many birds seem remarkably intelligent, including some jays, magpies, and crows.[18] Such "corvid" birds attest to the fact that, while some minimal brain size is required, it needn't be excessively large to house a pretty good little computational machine.

In keeping with this, certain facts suggest that the large brains of humans are overrated. To understand this comment, which may seem to go against conventional wisdom, think back to the previous chapter, where we summarized what we know about human origins and evolution. To recap, we "behaviorally modern" humans represent the most recent branch on a bushy tree of humanlike species. We eventually replaced everyone else, for unknown reasons. But the fossil record says that the expansion of the human brain size during evolution (measured by the space inside fossil skulls) began almost two million years ago and was completed several hundred thousand years ago—long before there was any archaeological evidence for behaviors of the kind we would identify with present-day humans. And while Neandertals (our closest extinct evolutionary cousins) had relatively larger brains than ours, they apparently never even made a simple drawing on a cave wall or created a bow and arrow, despite having opportunities to do so for more than two hundred thousand years. Also, there are babies born with major brain defects (or instances in which a large part of the brain is removed by surgery early in life, along with a tumor) who nevertheless end up with normal intelligence.[19] All this indicates that while the rapidly expanding human brain was apparently *necessary* for the earlier evolution of our human ancestors, it was not *sufficient* to take us to the current level of human intelligence. Michael Gazzaniga puts it another way: "Most of the evolution of the human brain, the presumed anatomy of intelligence, had occurred prior to any evidence for technological sophistication and, as a consequence, it appears unlikely that technology itself played a central role in the evolution of this impressive human ability."[20] This is one of the difficult conundrums we face in trying to understand the evolution of the human mind, and it suggests that *something else* happened *after* the attainment of maximum human brain size during evolution.

Regardless of brain size issues, there are quite a few species that seem to be primed to continue selecting for increased intelligence. A priori, all the mammals and birds listed in the preceding paragraphs would appear to have the requisite lifestyles to make increased intelligence a useful asset for species survival and success. Evidence even indicates that some dinosaurs (who went extinct about sixty-five million years ago) may have been warm-blooded, and possibly lived in groups. Yet—and this is the key question—since the days of the dinosaurs, only humans have become really, really smart. Why? Was there something special about humans that drove us to previously unexplored heights in our ability to use our minds? Was there an unlikely series of genetic changes that was required to permit our brains to develop our remarkable mental capability? Such questions are asked by those looking for an explanation of why our brain functions are so different from those of our nearest relatives. What other kinds of ideas have emerged to explain this?

All social animals have methods for communicating with one another. And some would say that many animals have unusual and complex communication abilities, and even some rudimentary form of "language." But no other animal's communication system comes anywhere close to having the range, complexity, and abilities of human language.[21] So even if a dolphin were smart enough to conceive of *Moby-Dick* (and if someone designed a keyboard for its flippers or beak), the animal wouldn't have the vocabulary to accommodate the rich detail that this tome would require. In this context, it has been suggested that the development of complex language provided the necessary selective force to evolve an organ with the cognitive abilities of the human brain. But this is just one plausible argument about the origins of human mental abilities. It could instead be that complex language evolved from some other ability of the human brain that preceded it.

This "language-drives-intelligence" notion is actually a hard one to conclusively prove or disprove (as are some other evolutionary scenarios). A major problem with understanding the origin of human language is that it left behind no "fossils" of any kind—until the advent of writing, which is a very recent human invention, apparently only about six thousand years old. All normal human children today can learn one or more languages easily. Thus the origin of our linguistic abilities must predate the common origin of all present-day humans. Prior to that time it is hard to know for sure. But evidence for a newly altered structure of the voice box (the larynx) and the throat (pharynx) in both early humans and Neandertals suggests that the ability to control sound might have been shared, at least in a somewhat similar manner.[22] In keeping with this, a human-specific change in the gene FOXP2 (which may contribute to our ability for speech production)[23] is also found in the equivalent gene of Neandertals and in our more recently discovered extinct cousins the Denisovans. Beyond such anatomical and genetic clues, it is hard to know when humanlike language emerged, and whether it was spoken or gestural (or both) in the beginning. It also becomes a chicken-and-egg question—which came first? Was there a gradual ramping up of intelligence and the use of that intelligence to create more complex language? Or was there some identifiable point, a "singularity," at which language emerged, providing a new driving force for the development of our relatively massive intelligence?

Any discussion of language must also consider that complex communication, with signals that are interpreted differently depending on the context, includes more than auditory processing of verbalizations. Studies of our primate cousins suggest that gestures and human-provided symbols can constitute a significant vocabulary, both alone and in combination with vocalizations.[24] Of course, most such popular claims for language in great apes are limited to a few individual animals cared for and evaluated only by one or two investigators in each instance. So there are difficulties with independent verification. But even if one accepts *all* the claims, the sum total of the evidence indicates that great apes do not have anything approaching a human vocabulary at a gestural or vocal level. It remains possible that we are not yet approaching the question in the optimal manner, and because of this we are unable to fully appreciate the communication abilities of other animals (especially marine mammals such as dolphins and whales).[25] Also, while we tend to focus on spoken words in thinking about language, the early development of this ability could have been built on a more complex, multimodal landscape,[26] involving, for example, sign language. Regardless, there is no reason to suppose that language itself was the defining selective force for human intelligence. Indeed, as we shall discuss later, there are humans with very limited language abilities who nevertheless have remarkable mental skills, including some with autism spectrum disorders. So while language is the most obvious, sophisticated, and universal manifestation of the complex mental abilities of humans, it may well be a consequence, not a driving-force cause, of our intelligence. In any case, whether or not one wants to buy into the notion of the primacy of language, this is not key to the essential ideas in the central argument of this book.

While language development could have been an important force in the positive selection for intelligence, similar forces should have been pushing numerous other social animals to become very smart as well. Smart animals should be capable of more complex communication than other animals. In social groups, improved ability to express knowledge, desires, and even ideas should be a good thing, as should be the ability to detect honest signaling of such expressions. Once animals developed communication systems as complex as those currently seen in elephants, dolphins, and apes, natural selection should have made these systems even better. For example, in a herd of elephants, the smarter ones should be more successful at mating and thus should pass their genes on more efficiently. And over time, future generations of elephants should become even smarter. Since this should be true across the board, why did selection not drive humanlike intelligence in other animals? Was it the environment? Likely not, as some other smart animals shared the very same environments in Africa with our human ancestors for millions of years. All this brings us back to the same stark question: Why is it that only humans are able to carry out so many special mental functions that seem missing from all other highly intelligent species, some of which have been around for millions of years longer than we have? What event happened to humans that made communication so special?

And what about simple behaviors, such as the use of tools? Again, there is nothing uniquely human in tool use; it is actually found in many animals. In addition to primates, some other mammals and birds and even octopuses use tools. Sea otters float on their backs and smash oyster shells on rocks that they balance on their bellies, but we don't expect that any of them will soon evolve the ability to make a metal fishing hook, line, and rod. So tool use is a sign of a certain level of intelligence, but there is no evidence that it is unique to our lineage or that it was the critical driving force in the final derivation of humanness. Of course, while many mammals and birds have been shown to make and use tools, theirs do not even begin to approach the complexity of human tools.

Just consider the complexities of making a bow and arrow—never mind the space shuttle. So while it is unlikely that tool use per se was the critical step in the evolution of human cognition, complex tool building in early human ancestors may well have favored the ability to carry out sophisticated mental processes in a step-by-step fashion, skills that were later taken advantage of as language and other uniquely human abilities emerged.

With the advent of DNA sequencing on a massive scale (a person's genome can now be read in a week for one thousand dollars!), scientists have begun to identify specific genetic changes that correlate with the evolution of the human condition. The basic protein machines of humans and chimps are more than 99 percent identical, and even the chimp-human differences that are found in the amino acid building blocks of the proteins mostly seem to have no functional significance. About a decade ago there was a great hope that simply sequencing the genomes (the total DNA) of humans, chimpanzees, and other closely related evolutionary relatives would quickly reveal the major differences that comprise the genetic contributions to our uniqueness. Having found the first example of a genetic difference, I even got actively involved in advocating for this approach,[27] and then participated in the analysis of the chimpanzee genome.[28] Alas, the availability of this genome sequence did not prove to be such a major step forward in explaining what makes humans unique. The fond hope that we would find a few key genes that made humans different did not come to pass, and the catalog of differences is now already in the hundreds.[29] As with many major biological mysteries, the genetic issues involved in human uniqueness turn out to be far, far more complex than one might have imagined. Furthermore, unique differences between humans and chimpanzees were found not only in the amino acid sequences that make up proteins but also in regions of DNA without any previously known functions.[30] Moreover, it is becoming ever clearer that human uniqueness arises from a very complex interplay of multiple genes with multiple environmental factors. So, as we said earlier, considering nature (genes) without nurture (the environment) is not very useful.

But this has not prevented some people from making provocative claims. For example, scientists found that a gene called *MYH16* (for a protein involved in making jaw muscles) was eliminated in human ancestors. They calculated that this happened about two million years ago and theorized that smaller jaw muscles might in turn have altered forces elsewhere in the skull, permitting an increase in skull case size.[31] In other words, a jaw muscle mutation may have indirectly allowed us to grow bigger brains, and this in turn may have allowed us to become human. While this story received much media attention, no conclusive connection was made between the gene and the actual functions of jaw muscles. Also, there is as yet no evidence that the larger jaw muscles of apes are preventing expansion of their brains. Finally, questions have been raised about the timing of this mutation in relation to its purported effect on human brain expansion.[32] Meanwhile, over the course of my own research, our group found that a gene called *SIGLEC11* is uniquely expressed in the human brain on cells called microglia, which are known to affect the growth and survival of nerve cells and their connections.[33] Moreover, the protein coded by this gene has a uniquely human sequence and recognizes another molecule called polysialic acid, which is involved in brain plasticity (the ability to change over time). But we have resisted the temptation to stitch these facts together into a theory about human brain uniqueness. After all, it may later turn out to be yet another "just so story." (Rudyard Kipling's *Just So Stories for Little Children*[34] presented highly imaginary but superficially convincing stories for small children. Over time, the term "just so story" has become a code word in academia for a scientific claim that sounds good enough to be true but is in fact not supported by conclusive evidence.) Regardless of the styles of different scientists, these are just a few examples of the many human-specific genetic changes that are being discovered by researchers—which will, in aggregate, begin to reveal the way genes have contributed to human features, always in concert with the environmental feedback to the genes. But we don't think these kinds of genetic changes will prove to be as critical as the as-yet-undiscovered changes underlying our human ability to deny reality.

On a more global scale, the determination of complete base sequences (A, T, G, and C letters) of the DNA from multiple species has allowed near-complete comparisons of our genetic information with that of several other animals. With these total genetic sequences in hand, it is possible, using computers, to search for regions of our chromosomes that have changed especially rapidly in the lineage that led directly to humans, including regions that do not carry information directly coding for the proteins. This has permitted the discovery of multiple regions of our chromosomes that may help define humanness with regard to our genetic makeup. Some of these genes are specifically turned on during brain development in humans,[35] consistent with the notion that such differences are important for promoting greater intelligence. However, none of these genes has yet emerged as the major explanation for the unique differences between humans and animals. Indeed, with the passage of time it is becoming quite unlikely that any single gene or region of the genome will be proven to have a dominant role. In retrospect this should not be surprising, given the long and complicated story of human evolution, which went through many twists and turns before finally giving rise to us, in very recent geological time.

In fact, many of the uniquely human genomic characteristics do not even occur in regions containing information coding for the structure of our proteins—rather, the changes are in DNA regions that affect how much or when a protein is made. This fits some earlier hypotheses that the primary difference between humans and chimps lies not in our proteins themselves but in the control (regulation) of when and where we make (express) those proteins. We refer here to the classic work of Mary-Claire King, who studied with Allan Wilson when she was a graduate student in the 1970s. The goal of King's thesis work was to find specific differences in protein or gene sequences between humans and chimpanzees. Back in those days such sequencing was very difficult. King persevered, but then found no major differences using the technology available at the time. What first appeared to be a failure of her thesis project was in fact later reported to be a landmark discovery—that human and chimpanzee proteins were so similar that it was hard to tell them apart.[36] This gave rise to the idea that most of the biological differences between humans and chimpanzees must arise from differential expression (different rates of production) of the same proteins. This hypothesis remains influential, but it is only partly correct. Over the last decade scientists have uncovered evidence of human-specific genetic changes of every other possible type, including direct changes to genes themselves and changes in the sequence of resulting proteins as well as wholesale deletion of genes. This seems to be another instance in biology in which all the theories about a given topic are correct in principle, but each is only partly so in practice. A 2012 review I coauthored lists more than two hundred uniquely human genetic elements. Details about these elements can be found at the Matrix of Comparative Anthropogeny (MOCA),[37] run by the CARTA organization. For those interested, MOCA also lists many other uniquely human differences between humans and the great apes, ranging from the anatomical and biochemical to the cultural and cognitive.

But when it comes to the scientific approach to explaining humanness, there is an additional issue to consider. It is hard to find a human who is uninterested in knowing where she or he came from, whether from an evolutionary, religious, or other perspective. So the difficulty for the scientist is twofold. First, one is asking a personal question with a built-in risk of bias—"Where did I come from?" In addition, the scientific community and the public are hungry for any evidence bearing on this issue. Thus there is a temptation for otherwise conservative scientists to examine indefinite findings and make more out of them than they really should. Indeed, a level of speculation is permitted in writings on human evolution that would not be tolerated in many other scientific fields (this book is an example!). Of course, given the difficulties of trying to understand human evolution, it is perhaps a good thing that so much speculation is allowed, as it stimulates thinking—but when one does speculate, it should be clearly disclosed.

Natural selection during human evolution also seems to have benefited greatly from changes in the timing of some growth events. Much of our growth (including that of our brains) is rather delayed relative to the timing of our birth.

This may well be one of the most important differences between humans and our closest evolutionary cousins—the delayed and protracted phases through which we develop. We are born early and then remain helpless for a long time.[38] One of the strong theories regarding human uniqueness has to do with this recent change in the life history, or ontogeny, of our species.[39] Thus, unlike our closest living evolutionary relatives, the great apes, most of our brain development occurs outside the womb, under the influence of our environment—which includes exposure to other humans and their languages, ideas, and cultural practices.[40] And because so many of our neural connections are generated and pruned after we have entered the real world, we can couple the modification of these connections to learning experiences that also occur outside the womb. Of course, this requires that we have parents who are willing to sacrifice their energies (and sleep) for a prolonged postbirth period, during which we are pretty much incapable of doing anything for ourselves. And grandmothers may have had a major role as well, a topic that we will return to later.

While humans seem to be an extreme case of delayed postnatal development, many of our primate relatives also make a significant parental investment in their young. But in the great majority of other primates, the mother is solely responsible for the care and provisioning of her infant prior to his or her nutritional independence. The exceptions to this general rule are called cooperative breeders—species in which the father and/or other members of the group participate in infant care. As emphasized by Sarah Hrdy in *Mothers and Others*,[41] humans are unusual in this respect. Some distantly related monkey species do display cooperative breeding, and the "lesser apes" (siamangs and gibbons) show joint paternal and maternal care. But the norm among all the great apes is that care of the infant is almost exclusively undertaken by the mother—with little interference (though much interest) from other group members. Thus humans are unusual compared to our closest cousins in having switched to becoming cooperative breeders. Hrdy suggests that, combined with helplessness of the young, cooperative breeding led to a situation in which it was necessary for the baby to monitor, communicate with, and solicit help from others, as well as for adult individuals in the group to communicate with each other, to assure survival of the baby—and the species. This would require not only self-awareness by individuals but also the ability to attribute mental states to others, a so-called theory of mind, which we will return to later.

Returning to genetics, there is no doubt that some of these newly identified genetic changes may have helped us become smarter. However, there is little reason to believe that any of them were the sole critical events in our evolution. Various types of genetic variants are arising on a regular basis in populations of creatures all over the globe. On the other hand, our evolution into superintelligent animals was a unique event in the earth's history. To put this in perspective, let's look back briefly at the history of multicellular animal life on earth. It is widely accepted that the basic body plans of all existing animals (and of animal types that have become extinct) were formed more than five hundred million years ago, mostly during what is known as the Cambrian Explosion.[42] This was not an explosion in the sense of an impact, such as that of a comet or asteroid, like the one that killed off the dinosaurs hundreds of millions of years later. Rather, it was an explosion of diversity in the types of animals that were evolving over a span of at least twenty million years. This may seem like a long time for an "explosion," but in evolutionary time it resulted in a very rapid and extreme change in the types of creatures that were evolving. We had the emergence of animals with hard cuticular external skeletons (such as insects), the beginnings of animals with backbones (including us), and many other types. The important point for us here is that all the basic types of animal body plans we see currently appeared around this time and have been in existence for many hundreds of millions of years.

Since the Cambrian Explosion, we have seen each type of animal undergo extensive refinements and, most important, diversification. For example, when we consider vertebrates (animals with backbones), we see stupefying panoplies of elaborations on the basic theme. If there is some selective advantage to having a long neck, animals with long necks will have higher fitness than those who don't, and future generations will, on average, be more like the animals with long necks. This will continue, and before long (in an evolutionary sense) giraffes will be walking the earth. What—you've evolved to live on land and now you've "changed your mind" and think you would be better off going back to living full-time in the water again? No worries; evolutionary selection can select for webbed feet, which will eventually morph into flippers, and you can become a whale or a dolphin (complete with a tail fluke, which has no counterpart in the terrestrial mammal body plan). Given enough time, evolutionary selection can also convert your nose into a blowhole, change your physiology so that you can hold your breath for a long time, and allow you to withstand the immense pressure changes that will permit deep diving.

Studies in developmental genetics are also uncovering the extremely intricate webs of gene and protein interactions that turn a simple early embryo (a nondescript ball of cells) into the magnificently elaborate organisms we see today, with exciting connections to evolution that are sometimes called Evo-Devo. This kind of work provides a strong molecular underpinning to the notion that there is enough genetic variation in a large population of animals for evolution to exaggerate virtually any structure or physiological process, as long as there is sufficient evolutionary selective pressure to do so. Moreover, depending on the population size and degree of selection, these changes can happen over a relatively small number of generations.

Of course, we must remember that we are measuring time in an evolutionary sense. Polar bears may not be able to change fast enough to escape from extinction in the face of global warming and the melting of the Arctic ice, but if the temperature changes occurred over tens of thousands of years instead of decades, they might have a chance. After all, recent evidence indicates that polar bears basically evolved from some northern brown bears that became trapped in the Arctic during the last ice ages. In other words, the distinctive appearance, behavior, and unique diets of polar bears may have emerged only in the last few hundred thousand years.[43] This is a demonstration of how rapidly natural selection can push evolutionary change when the environmental constraints are extreme. And if humans direct the selection through controlled breeding, it can move even faster—just think of all the varieties of dogs we have derived from a wolf ancestor in a few thousand years (and most present-day Western breeds were established just in the last two hundred years).[44]

A good example of how effectively selection can exaggerate a trait can be found in our senses—smell, sight, etc. If you take your dog for a walk in the neighborhood, it probably likes to stop to sniff many of the objects along the way. Your dog's nose is good enough to detect faint residual odors that may be weeks old—"seeing" the history of each bush and rock as told by the lingering odors of each previous dog that decided to mark it. Like many animals of its ilk, your dog has a sense of smell that is far better than yours. The keen vision of a hawk can pick out a tiny rodent scurrying through the brush from hundreds of feet in the air. An owl can hear the same rodent and, in the dark of night, use this ability to zero in and snatch it up. A viper "sees" the same rodent by using specialized pits in the front of its head to sense the infrared radiation of the mammal's warm body. Bats use high-pitched auditory signals to detect and capture flying insects in the dark. Using its ability to detect a wide range of vocal frequencies, a whale can hear a friend call from half an ocean away. This list could go on and on. The point is, these extreme sensory abilities have evolved numerous times through many, many different lineages.

Ironically, it appears that when it comes to the senses, we humans seem to be mostly going toward deteriorating function. Our senses of smell and hearing appear reduced from what they were eons ago, and some genetic explanations are available (loss of genes for smell-detector proteins).[45] This has likely happened because we don't rely on these senses so much anymore. We have managed to maintain tricolor stereoscopic visual abilities inherited from our primate ancestors, but even this sense does not seem to be any more acute than that of other primates. Further studies of the sense of touch are required to know whether or not humans have improved in that realm. But overall, there does not appear to be anything special about our human senses.

Anyway, the degree of genetic variation in a population does not seem to be much of a factor limiting the evolution of extreme sensory abilities. If there is selective pressure to elaborate and exaggerate a modality that already exists, the genetic tool box seems to have plenty of options for getting one to the desired place. And it does not require hundreds of millions of years of gradual evolution. Why should we expect brains to be any different? Mammalian brains have not changed too much in terms of the types of structures present or their overall organization. Our brains are relatively larger than those of most other animals. But in terms of basic structure, we so far do not seem markedly different from our primate relatives. And, as discussed earlier, our brains are obviously not as large as those of an elephant or a whale. It is only that our brains are somewhat larger relative to our body size. Making a human brain may have required elaboration and/or exaggeration of existing parts, but there is no reason to think that any unique or difficult genetic events needed to occur along the way.

While it is naive to view increased brain size as *the* defining event in human evolution, an enduring mystery is why the human brain continued to expand during our evolution between two million and three hundred thousand years ago, despite the increasing difficulty of delivering newborns with large heads through the narrow human pelvis[46]—a process that (before modern obstetrics) caused the deaths of many mothers and infants. Although theories abound, we really do not know what the driving force was for the expansion of the brain in this early period of evolution. This is especially so because, as we said earlier, the archaeological record over this time period shows no indications of a major change in intelligence (no marked improvement in tool making, for example). Regardless, it is hard to imagine that inability to evolve a larger cranium has been the major stumbling block in the development of chimpanzee intelligence. Given enough time, making an existing part better has never been much of a challenge for evolution, at least not from the standpoint of the available genetic tool box.

So there are lots of reasons why being intelligent can be a good thing. This is especially true in animals that hunt collaboratively and live in stable social groups, where competition with one's group mates is likely to be of particular importance. And once being smart becomes a weapon in the fitness battle, being a little smarter is going to be a little better. This should drive an intelligence arms race. There is no reason to suggest that the level of genetic heterogeneity in populations is a significant limiting factor in becoming very smart, at least over the time scale during which vertebrate brain evolution has occurred.

Much research and theorizing has focused on asking what unique set of factors provided the positive driving force for human intelligence and what unusual genetic events (mutations) had to occur to allow us to make brains that could carry out our unusual thinking abilities. But these may be the wrong questions. We think that nothing uniquely special was required to push the development of our unusual brains. Being smart has advantages, and there are no molecular constraints on the evolution of bigger or more efficient calculating machines. Our question, then, is not what rare events had to happen to drive the special evolutionary elaboration of human minds. Rather, the questions that should be asked are *Why did it take so long* and *Why did it happen only once?*

The reasonable answer is that there are also costs associated with being too smart. Just as the tail of the peacock makes it less than the most graceful of flyers, there is a selective price to pay when our brain function is exaggerated to an extreme extent. In order to become smarter, the advantages to fitness, which are fairly straightforward, have to outweigh the disadvantages, which are less obvious. So what we are proposing is that there was a very difficult *psychological evolutionary barrier* that had to be crossed before any further development of humanlike intelligence could occur.

In examining the potential costs of being smart, we will suggest a surprising answer to the question of why it was so difficult to evolve a species that included individuals who could create a Taj Mahal or put a man on the moon. We will also uncover the disturbing reason why we are inclined toward such self-destructive endeavors as building nuclear weapons and filling the atmosphere with dangerous greenhouse gases. We will ask you to consider an idea that at first glance seems counterintuitive, and may well be initially dismissed by some experts. But remember that the history of science is full of counterintuitive ideas that initially generated negative responses from experts. It is only in retrospect that such ideas become "obvious." A classic example is Alfred Wegener's theory of continental drift (currently called plate tectonics),[47] which presented the now obvious fact that the surface of the earth consists of moving plates, so that continents have been forming and breaking up over time. Partly because he was not a card-carrying geologist, Wegener's ideas were completely rejected by the established experts of the day, even though the theory is now considered a cornerstone of geology. Further back in time, we have the Polish astronomer Nicolaus Copernicus, who went against the orthodox view of the day (that the earth was the center of the universe), claiming instead that the earth revolved around the sun—the heliocentric view that we now know to be correct. So it is a good thing to keep an open mind about new ideas, even if they go against orthodoxy. But Alfred Wegener actually died during an expedition to seek further evidence for his own theory, before it became accepted. And Copernicus was on his deathbed the day that he was shown an advance print copy of his epochal *On the Revolutions of the Heavenly Spheres*. He, too, did not live to see his unorthodox views vindicated. So if you are going to go up against orthodox views with a radical new theory, it is preferable to stay alive for a while after you publish it.

CHAPTER 3

There Are No Free Lunches or Free Smarts

We are only now beginning to acquire reliable material for welding together the sum total of all that is known into a whole; but, on the other hand, it has become next to impossible for a single mind fully to command more than a small specialized portion of it. I can see no other escape from this dilemma . . . than that some of us should venture to embark on a synthesis of facts and theories, albeit with second-hand and incomplete knowledge of some of them—and at the risk of making fools of ourselves.

—Erwin Schrödinger, in *What Is Life?*

Extreme exaggeration of a specific trait within a species can confer a fitness advantage under specific situations. As we said, the long neck of the giraffe allows it to browse high in the trees, and the peacock's outrageous tail makes it more attractive to peahens. However, in each such case, there is a selective cost, a price to pay, for a trait that is amplified. In other words, it is likely that there were trade-offs for any unusual trait that evolved in only one or two species— trade-offs that made such a trait rare in the first place. Giraffes' long necks make them less agile, and the animals are particularly vulnerable while they're reaching down to drink water. The bull moose's enormous rack of antlers must be carried through the forest day and night, just so that he will be better able to acquit himself in the risky combat for mating privileges. And even one-celled bacteria experience trade-offs: A mutation that confers resistance to an antibiotic can sometimes make the bacteria less fit if the antibiotic is not present. In all these cases, and countless more, there is a negative effect on fitness with the extreme expression of a particular trait. If the sum of positive and negative effects comes out on the plus side of the ledger with regard to reproductive success, then a trait will persist. But as soon as the negatives outweigh the positives, continued elaboration of the particular trait(s) will cease, because the individuals with those genes will not procreate enough. There will be variations around this point, but in the population as a whole, a stable average will be attained.

So this raises the question: What are the downsides to being smart? (And by "smart," we don't necessarily mean human smart, just smarter than the average monkey.) While everyone agrees that there are costs associated with other unusual evolutionary advantages, there is less discussion of the costs that might result from becoming "too" intelligent. Given the apparent uniqueness of some of the unusual mental abilities of humans, there must have been one or more major barriers against achieving this state. Before diving into this question, though, it might be useful to look at a similar but slightly simpler question.

As we mentioned in the previous chapter, specific senses are developed to extraordinary degrees in various animals. Indeed, some senses have been developed to extremes many times in many lineages when it suited the needs of natural se-lection. Dogs, or even vertebrates as a whole, are not the only animals with an uncanny sense of smell. Certain moths are extremely sensitive to specific odors, such as the molecules that they use to home in on a potential mate. Indeed, even a one-celled bacterium can sense and count single molecules that impinge on its surface and can use this sense to make de-cisions as to which way to swim based on the frequency with which it encounters the stimulant.

So if it's relatively easy for evolution to generate such ex-tremely sensitive sensory prowess, how come we don't have lots of animals that can simultaneously smell as well as a bloodhound, see as well as a hawk, and hear as well as an owl? Heck, why not throw in the ability to see infrared radi-ation, like a pit viper; to detect sonar signals, like a bat; or to see ultraviolet light, like some birds? This becomes even more vexing when we note that many animals seem to have actu-ally lost sensory abilities over time. In the desert southwest of the United States, we have peccaries with a very acute sense of smell, which allows them to root out food. However, they see very poorly. There is no purely genetic reason that they cannot also have eyes that function at least as well as, say, the deer that live in the same habitat. What possible advantage could there be to being partly blind? Closer to home, it is in-teresting that when we compare humans to chimpanzees (our closest living evolutionary cousins) we find very few features of ours that can be considered "superior." In fact, one can only think of three. The first is our obviously greater mental ability. The second is our capacity for long-distance running. And the third, possibly, is the anatomy of our hands, which gives us greater control over our thumbs and fingers.[1] In al-most every other way we actually seem inferior: We have weaker senses; poorer muscle strength; tender, furless skin; our childbirth is difficult and dangerous; and our young are almost completely helpless, among many other deficiencies. So why did humans become inferior in so many respects, even while our cognitive abilities were improving? Could it be that the powerful abilities arising from human culture and learning have made genetic selection less important? This might even have allowed our genetically hardwired abilities to deteriorate (a phenomenon called relaxed selection) without serious consequences to the survival of our species. This deterioration may have even been converted into an advantage: It allows unusual human individuals (and their unusual and useful genes) to survive—individuals who might not have survived in another species.[2]

Regarding the more general question about limitations in senses, a plausible explanation is that each species "learns" (in an evolutionary sense) to focus on the sensory inputs that are doing the most good, from a fitness perspective. Too much information can lead to some level of mental confusion, and so one needs to ignore those sensory packets that are going to distract from the main goal. Evolution could perhaps have dealt with this by altering the way that sensory data are processed in the higher centers of the brain. Or it could just limit the processing of less useful data by "executive control." The importance of focusing on the information that matters most is clearly evident in the photoreceptors of animals and the neurons that process their information. For example, in some frogs it has been shown that the nerves are connected in such a way as to make them especially sensitive to motions like those of a fly rapidly moving across the visual field.[3] No one has proved this, but this wiring would probably make the frog less capable of slowly scanning the words on a page. But for a frog (except maybe for Kermit, the lovable character in *Sesame Street*), the cost of not being a speed-reader is small indeed, if it means you can catch more bugs.

As evolution for greater cognitive function proceeds, we can expect similar issues to crop up. If an animal tries to compute all the interconnected possibilities before reaching a conclusion, confusion can result. At the very least, simple decisions can become overly time-consuming and even risky. Imagine an antelope on the Serengeti Plain in East Africa. There is a snapping sound in the tall grass nearby. The antelope could carefully consider all the possibilities. Possibility number one may be that this resulted from a branch falling from a tree. How close is the nearest tree to the supposed source of the sound? What may have caused the branch to fall? Was it just because the branch was dead and the wind was blowing? Is the wind blowing? By the time all the possibilities have been identified, probabilities assessed, and potential repercussions analyzed, the antelope may be in the jaws of the lion that stepped on the twig in the tall grass. Thinking too much before making a decision can be fatal. This will in turn have a negative effect on fitness. This is a relatively simple example of how one has to balance the benefits of new mental abilities with possible negative outcomes. And in this instance the evolutionary process is simply "searching" for an optimum state rather than trying to cross a major, nearly insurmountable barrier. This doesn't mean that better computational power is necessarily a drawback. It only points out that it must be managed properly. This simple example shows that in many instances it is desirable to have a short circuit that can override raw analytical tendencies. The antelope might be better off to just jump and think about it later, if at all. This line of thinking echoes

Daniel Kahneman's Nobel Prize—winning work, which shows that we have two systems driving the way we think. While system 1 is fast, emotional, and intuitive, system 2 is slower, more logical, and more deliberative. Counterintuitively, system 1 tends to dominate most of our actions and decisions.[4]

Indeed, evolution seems to have "decided" that in many circumstances it's best not to think at all. It's not only our legs and arms that react spontaneously to things happening around us; we have "short circuits" in our brains (and hormones) as well. Some of these correspond to what we call emotions. What is being "happy," "sad," "fearful," "anxious," or "angry"? These are not rational conclusions we reach based on a detailed assessment of the pluses and minuses of a given situation and a careful computation of the response that is in our best interest.[5] They are mental and physiological responses to particular sets of circumstances in which evolution has determined that a given response will, on average, be best in maximizing our fitness. Emotions have been shaped by selection over countless generations to guide our behaviors in the absence of exhaustive and perhaps futile analysis of situations. They are, quite literally, irrational.

So how is it that our emotions, which seem to regularly get us into trouble, maximize fitness? To answer this, we need to consider some caveats. The first is that no triggered behavioral response is always going to be the optimal choice in every situation. The antelope that bolts at the slightest disturbance may in fact run directly to a silent stalker, who may have been noticed if the antelope had taken a more measured assessment of all the options before running. However, most of the time, a quick reaction to the known potential threat will be the best choice, especially in a species with many predators. Averaged out over many instances, the delay will be more costly than the potential for going the wrong way. Selection is a very good calculator, and a success difference of a fraction of a percent can be meaningful when applied over millions of trials.

For humans, a knee-jerk response triggered by emotions or other feelings is even more likely to be a mistake today. This is because of the previously discussed disparity between biological and social evolution. We live in a world that is radically different from the one for which we were evolved. A suppressed fear response that may have served us well in our camps and home bases, where we were surrounded by trusted longtime companions, may not be appropriate in a twenty-first-century metropolis. We are also constantly fighting innate tendencies that are not healthy for us. For example, for a subsistence hunter-gatherer, it is generally not a bad strategy to eat all you can when you are lucky enough to find a food rich in energy sources (sugar or fat). This may be why we are "designed" by selection to find these foods so pleasant. However, for some of us fortunate enough to be surrounded by essentially limitless quantities of sweets and greasy deep-fried foods, it is decidedly unhealthy to continuously gorge ourselves. The irrationally driven behavior that was good for us thousands of years ago is not in our best interests as citizens of the affluent metropolitan society of today. In fact, the vast majority of what most of us do in so-called developed societies has very little to do with what our bodies were evolved and likely optimized for by earlier evolutionary selection.[6] This environmental "mismatch" negatively affects not only our nutrition but also our physical strength and cardiovascular conditioning. In all these instances, our mental abilities, manifested in the form of "culture," seem to have overshadowed our innate genetic mechanisms. As we suggested earlier, it is even possible that supplanting our built-in abilities with culture may have allowed the loss or degradation of some innate genetic endowments, such as our senses of smell and hearing.

Emotions are just one example of how our brains have evolved to take shortcuts, to try to generate and reinforce somewhat optimal behavior in a reasonable amount of time. During animal evolution, there must have been a general need to continuously balance raw computational power with speed and other factors that allowed us to make the best decisions. An analogy can be found in computer programs that are designed to play chess. A chess player could, in theory, respond to any opponent's move by computing all possible responses through to the end of the game. But because of the enormous number of potential directions that the game may take, this is an impossible task for a human, and so the best players learn strategic guidelines to follow in the placement of their pieces. This strategic knowledge is combined with the tactical ability to visualize all the reasonable possible outcomes for a certain number of potential moves into the future. The very best players have a firm understanding of these strategic concepts as well as an ability to see specific possibilities far into the future. Chess-playing computer programs for many years used a similar strategy. Human strategic thinking was programmed into the machine and combined with standard computational and tactical calculating abilities. These programs could play excellent chess, but until recently did not beat the very top humans. As computational power increased, though, programmers were able to shift the balance in favor of comprehensive computations of all potential series of moves into the future, involving relatively little human-type strategic thinking. The best of these programs outperform the older versions and can beat even the top grand masters. Is it possible that human brains could eventually have such massive computational ability? Could we evolve into biological computers on the same scale? In fact, we may already be much closer to that point than you think. But it may not be a very good strategy for success in life, unless success is measured only by your ability to do one specific task (such as playing chess) extraordinarily well.

Ever since the classic 1988 movie *Rain Man*, many of us have become familiar with what is called the savant syndrome. This syndrome typically combines an extraordinary mental ability with some form of mental disability,[7] sometimes autism[8] (*autism* is actually a generic term for a range of mental conditions called autism spectrum disorders, which we will discuss later). The character Raymond Babbitt in *Rain Man* was based on a real person who memorized ten thousand books and had detailed recall of a bewildering array of information in areas such as sports, geography, and history, among others. He could read extremely rapidly, and had the ability to read the left and right pages of a book simultaneously. And all this despite the fact that he had a low measured IQ, required help to take care of his personal needs, and related very poorly to others. Some other examples illustrate the scope of the computational potential of the human mind and also begin to illustrate some of its limitations. One of the earliest described savants was Thomas Fuller, as reported by Benjamin Rush in 1789. In ninety seconds Mr. Fuller correctly calculated the number of seconds lived by a man aged seventy years, seventeen days, and twelve hours (2,210,500,800), even correcting for leap years. But according to Rush, he "could comprehend scarcely anything, either theoretical or practical, more complex than counting." There is a pair of identical twins who can calculate the calendars covering more than forty thousand years and who remember the weather for every day of their adult lives. They also can compute twenty-digit prime numbers.[9] But they are stumped by otherwise simple arithmetic problems. Many other savants with extraordinary abilities are known from around the world and from historical records. Their individual abilities may lie in areas such as music, art, mathematics, or spatial skills. One thing they have in common is that their abilities typically involve remarkable memory and recall. Also, while their skills may be extremely deep, they are usually narrow in scope.

There are many theories about the savant syndrome, and several of them—not just one—are likely correct. The unifying feature appears to be that major aspects of normal human brain abilities have been diminished, either by a physical cause, such as brain damage or surgery, or, for unknown reasons, such as an autism spectrum disorder. The human brain in such situations is able to focus the remaining functional components on a few very specific abilities. This is probably reflective of the remarkable plasticity of the human brain. In other words, such individuals appear to adapt additional portions of the brain to take over functions that would otherwise not be used or, conversely, to use otherwise unused parts of the brain for functions for which they were not originally intended. Extreme savants, like those described above, are relatively rare; savants whose abilities and deficits are less extreme are much more common.

What has savant syndrome got to do with "normal" people? Unfortunately, there is no comprehensive explanation for what causes it. But its association with other mental disabilities has led some to suspect that savant syndrome is not so much an instance of gaining fantastic powers of memory or calculation but rather a case in which the latent abilities that we all possess have been uncovered. That is, our brains are probably much better calculators than we know, with abilities hidden by layers of complicated cognition that suppress our awareness of them. Additional support for this idea comes from cases in which savant-like abilities are sometimes unleashed in persons who suffer a brain injury.[10] Even more suggestive are cases of previously normal persons who acquire new and prodigious artistic skills following the onset of certain kinds of dementia.[11] In these cases of injury or dementia, individuals with normally functioning minds unlocked exceptional abilities following damage to a part of the brain that reduced other normal functions.

Perhaps equally remarkable are the rare individuals (around twenty known cases, all of whom appear otherwise mostly normal) who can recall events of most days of their lives in nearly perfect detail (one of them is the actress Marilu Henner, whom you may remember from the hit TV series *Taxi*). Far from being a benefit, this kind of "superior autobiographical memory" (technically called hyperthymesia)[12] is seen by these individuals as a difficulty, especially with regard to personal relationships. How would you like to have a spouse who turned out to be right every time she or he disagreed with your recollection of something? Not surprisingly, these people typically do not have long-term stable marriages! Hyperthymestic individuals may also have a below-average memory for arbitrary information that is not autobiographical, another price to pay for this ability.

All this is strong evidence that we are walking around with calculators that are far more capable than we can imagine. On the other hand, the phenomenology of savant syndrome fits very well with the general notion that too much informational processing can be deleterious to overall functioning and fitness. Our brains are extraordinary calculating machines, if we allow them to become obsessively focused on a particular task. But this level of focus compromises our ability to effectively process all the relevant information that any life situation presents. So as animals became smarter, they also evolved many filters that make "executive" decisions as to what should command attention. Humans, with our exceptionally powerful brains (especially the prefrontal cortex),[13] have also become extremely good at generalizations. We deal with life in ways similar to the way a strategic thinker plays chess—making moves that we "know" are good not because we are calculating each specific course but because we are unconsciously drawing conclusions from emotions, similar experiences, and other global patterns. We may be using a fair amount of our calculating ability, but it is not necessarily close enough to the surface for us to be aware of it. On the other hand, savants are not distracted by all those executive functions of the brain, and this perhaps allows them to access the deeper cognitive calculator and focus it on a specific task.

So in order for an animal species to become evolutionarily smarter and smarter, it would need to develop better executive brain functions. It would need to get better at generalizations and at prioritizing the information and processing the events that will occupy its immediate attention. One could think of savants as individuals who have lost these evolved executive functions to one degree or another, allowing their innate, exquisite calculators to obsessively usurp their attentions. For example, in her fascinating book *Animals in Translation*,[14] Temple Grandin proposes that some autistic people understand the minds of our animal relatives better than the minds of other people. Dr. Grandin is a remarkable individual. She is herself autistic, but has adjusted to understanding "typical" humans by using tricks like visual thinking.[15] She has also made herself an authority on domestic animal behavior. Grandin contends that being autistic provides her with unusual insight into the workings of her subjects, primarily livestock. She argues that animals are much more likely to be distracted by inputs that most humans would ignore, largely because of our enhanced abilities to generalize and conceptualize. But, perhaps, is it also because of our ability to deny or ignore reality when we consciously or unconsciously choose to do so?

The ability of humans to subjugate our immense brainpower is not unique; it's just exaggerated to match the scale of our overall high level of intelligence. Over time, smart animals would be expected to develop compensatory mechanisms commensurate with their need to keep from being confused or obsessed by the complexities of their thoughts. We can see these mechanisms in other animals now. Most vertebrates with whom we are familiar have emotions of the most basic sort—fear, anxiety, etc. Many mammals clearly have nurturing emotions that could pass for love (and that are not just directed toward progeny), and anyone who has watched otters having fun or who has played fetch with a dog must detect signs of happiness.

So being very smart can be useful, but it comes with costs. The primary cost is potential confusion from too much analysis, or obsessions that result from focusing too deeply on a particular part of an immense calculator. However, it appears that as animals got smarter, evolution was able to keep the costs down by developing compensatory "executive" brain functions that maintain tighter control of previously automated functions. Animals have been able to maximize their cognitive abilities by proper organization and conceptualization. These two characteristics—increased computational ability and more developed executive functions—can evolve in small steps together, as needed. Thus these costs don't seem to have provided significant handicaps for continual cognitive advancements. Humans manage pretty successfully with minds that are very far advanced compared to other animals. One would think that these issues should not have been very problematic for the evolution of ever-smarter elephants or dolphins.

So mental confusion resulting from complex thinking and obsession arising from too much attention to detail do not appear to be critical handicaps that should have inhibited the continued development of progressively smarter animals. However, there is *one additional major hurdle* that presents new challenges for intelligent beings. Unlike the continuous relationship between the benefits of increased computational power and the cost of confusion or obsession, this hurdle presented *a discrete, novel barrier at a specific point in our evolution, which required a totally new strategy in order to overcome it*. It is one thing to be self-aware. But it is quite another to also become fully aware of the self-awareness of others. This may sound like a positive step in the evolution of the mind. Paradoxically, though, it may not be so at the outset, because of negative psychological consequences, which we will discuss later. It is actually a nearly insurmountable barrier, which only we humans have managed to break through, aided by a most unusual psychological evolutionary trick. But before we move on to consider this nearly insurmountable barrier, we must consider the many levels of awareness that human and other animals can experience.

CHAPTER 4

Many Levels of Awareness

Blushing is the most peculiar and most human of all expressions.

—Charles Darwin, in *The Expression of the Emotions in Man and Animals*

Man is the only animal that blushes—or needs to.

—Mark Twain, in *Following the Equator*

Throughout the rest of the book we will continue to use the term *self-awareness* to mean the knowledge of one's own self as an individual, or knowledge of one's own personhood. This is, of course, not a precisely definable thing but rather a state of mind. One type of experimental evidence that may denote self-awareness of the kind we are talking about is the mirror self-recognition test, adapted by Gordon G. Gallup Jr. to evaluate self-directed behaviors of animals that emerged in front of mirrors.[1] In this test, an animal is first allowed to get accustomed to a mirror (so that the reflection does not seem threatening). Some animals will then attempt to interact socially with the reflection, thinking it is another individual of its own kind. The subject is then put to sleep and an innocuous spot is painted on an otherwise invisible place on its body, such as the forehead. Upon awakening and looking at the mirror again, most animals will not react any differently. However, animals that may be recognizing themselves in the mirror will show a much-increased interest in their reflection, and, if it is physically possible, they will try to touch the spot. This evidence, and the subsequent self-directed behavior of animals that seem to understand that the image reflects their own bodies, is used to indicate that the animal in question has self-recognition, or self-awareness.[2]

Skeptics, such as Danny Povinelli, have argued on purely logical grounds that mirror recognition may only be evidence for an integrated body image, not necessarily self-awareness. But body image is something that many animals probably have. As Derek Denton[3] points out, birds flying at high speed through the forest must have a pretty good cognition of their own body image and its dimensions—or else there would be a lot of dead birds on the forest floor. And at least in the case of chimpanzees, Denton has provided fascinating evidence that there is more to it. In one experiment, he presented a distorting circus mirror to chimps who had already passed the basic mirror test.[4] As Denton later wrote to me:

> These chimps had an earlier experience of viewing themselves in an ordinary mirror several times, had became aware that it was themselves, and examined the otherwise inaccessible parts of their body with the aid of the mirror. But when they were first exposed to the circus-distorting mirror, [with either] concave or convex distortion ... there was a small startle or surprise-like reaction. However, very soon they started to sway from side to side while watching intently and the impression was strong that they were establishing a contempo-

raneity of their willed movements with the movement of the image. That volitionally induced contemporaneity seemed to soon cause them to decide that the odd image was themselves. A distinguished scientist with whom I looked at the film has put [forward] the proposition [that] the chimpanzee was conducting an experiment when swaying and watching—i.e., elementary cause and effect!

Regardless of what you may think of its experimental value, some (but, curiously, not all) chimpanzees, orangutans, dolphins, orcas, elephants, and magpies seem to have passed the mirror test (the data are weak or limited in some instances).[5] Since the common ancestor of all these species lived tens of millions of years ago, and since most present-day species do not have this ability, it appears that some kind of basic self-awareness has *evolved independently* several times. Interestingly, human babies are not born with such self-awareness, but only achieve it at about two years of age.[6] In passing, it is noteworthy that while most chimps and orangutans seem to have such self-awareness, most gorillas have failed the mirror self-recognition test.[7] But gorillas are genetically closer to humans than we are to orangutans. So presuming that the shared common ancestor of humans, chimpanzees, gorillas, and orangutans already had this ability, gorillas apparently then lost it during their later evolution. Given the recent availability of the genomes of all these species, there is a chance that comparisons may reveal the genetic basis for this apparent loss of self-awareness in gorillas.

We have referred several times to a concept called theory of mind. This term roughly means that an animal is not only aware of his or her own individuality and intentionality[8] but also realizes that other individuals have their own minds and are thus independent "intentional agents" (this is sometimes also called intersubjectivity). Another way to put it is that an individual is not only self-aware but also knows that other individuals are self-aware. Still another way to put it is that an individual can attribute mental states to others. This advanced level of awareness of the self-awareness of others is sometimes also called second-order intentionality. Among humans, we can go further and have two individuals who mentally relate to each other while together attributing mental states to a third individual (sometimes called third-order intentionality).

While this general point may be clear, the amount of jargon may become befuddling to some readers. To make matters worse, experts do not agree on clear-cut definitions of many of these terms. This is partly because the terms actually represent *various stages along a continuum*. I present this continuum below in four *arbitrarily* designated stages. (For aficionados, some more technical terms, taken from a paper by Call and Tomasello, are presented in the endnotes.)[9]

Stage 1: Self-awareness. This term has many meanings, and to a layperson it could simply mean awareness of one's own body, something that most animals must have. But the term is used in scientific circles to imply something more—awareness of one's own personhood.[10] Most animals studied to date do not show clear-cut evidence of such self-awareness, at least if we use the criterion of being unable to recognize their own image in a mirror. But as we have said above, some animals and birds with self-awareness seem to be aware of their own personhood. This stage has also been called first-order intentionality.

Stage 2: Rudimentary theory of mind. In this stage, which is likely achieved by some chimpanzees and perhaps some other nonhuman animals, self-aware individuals have some realization that others can have their own perceptions and goals—that they are "intentional agents." But they still do not fully comprehend that other individuals are fully self-aware themselves; that is, that others have minds of their own. This stage might be called rudimentary second-order intentionality.

Stage 3: Full theory of mind. At this stage, which only humans seem to have achieved, one individual fully understands that another has an independent mind like his or her own. This might be thought of as a full awareness of the full self-awareness of others. At this point one also understands that other individuals can hold false beliefs. This stage might also be called full second-order intentionality. Having understood false beliefs on the part of others, it is also possible to hold a false belief oneself, even when that belief goes against one's own knowledge of the facts—this is known as self-deception, a common human failing.

Stage 4: Extended theory of minds. At this more advanced stage, two individuals not only attribute mental states to each other but also jointly understand that a third individual has a mind. This stage can be called third-order or multiorder intentionality. This stage seems to be achieved only by humans. Of course, these multiple orders of intentionality can also be present in the mind of a single person who is attributing mental states to several other individuals. Indeed, humans can have an understanding of the mind of someone they have never met, or even the mind of a fictitious person, such as the famous detective Sherlock Holmes.

Besides *intentionality*, there are many other terms used to describe aspects of this mental continuum—but there is no single term that explains it all. To add confusion, the Premack and Woodruff paper that is cited as the original source of the theory of mind seems to have used this term to address both self-awareness and the attribution of mental states to others.[11] Others later began to use *theory of mind* to refer only to the attribution of mental states. All things considered, the rest of this book will stay with the traditional term *theory of mind*, warts and all, and simply abbreviate it as ToM.[12] But to reemphasize, these arbitrary "stages" of ToM *represent a continuum, with no shining bright line between these different stages of awareness*. The point is that only humans seem to have attained a full and extended ToM. So the question is, What has kept other species with rudimentary ToM from going on to acquiring full ToM, despite tens of millions of years of opportunity?

Dog lovers are probably wondering why their favorite animals have yet to be mentioned. Most long-term dog owners would swear their pets must be mentally special because of their remarkable ability to relate to humans. In fact, Charles Darwin himself was greatly influenced by the dogs in his life.[13] However, tests so far do not show evidence of self-awareness in dogs. For example, they do not recognize themselves in a mirror. And one expert even opines that "domestication does not seem to have given dogs the ability to read our minds, or even to understand that humans are capable of independent thought."[14] Regardless, dogs seem to take cues from humans remarkably well. In fact, recent research has experimentally documented what dog owners already know—that our canine friends have the ability to follow and understand human behavior, intuitions, direction of pointing, and even direction of gaze.[15] Like us (and unlike most other primates),[16] dogs even have whites of the eyes, which makes it easier for us to follow each other's direction of gaze from a distance.[17] And much has been written about the unconditional love that dogs can have for their owners.[18]

In light of all this, it seems surprising that dogs have failed to recognize themselves in a mirror or show a sense of self-awareness. How can we explain this? One plausible account is that while dogs may have a rudimentary ToM in relation to their owners (and to humans in general), they do not have personal self-awareness. But how can this fit with our previous statement that self-awareness appeared first in evolution, followed by the crossing of a barrier to attain a full ToM? To understand this apparent contradiction, one must realize that dogs are not the product of *natural* selection but rather of *artificial* selection by humans. Available evidence strongly indicates that dogs are descended from gray wolves that were domesticated (one or more times) after the emergence of modern humans but before our final spread across the whole planet. While the exact dates of this domestication remain uncertain, the bottom line is that we humans domesticated dogs, and in the process artificially selected for these very abilities that we find so remarkable. In keeping with this, studies of recently tamed captive wolves suggest that this capacity to relate to humans is not a general innate ability of canines.[19] And while experimental attempts to domesticate Siberian foxes by selecting for nonaggressive behavior succeeded after many generations,[20] it does not appear to have yielded the kind of mental bond that dogs seem to have with humans. In contrast, dogs brought up with minimal human contact still have the ability to relate to humans. Thus the original domestication event(s) that gave rise to present-day domestic dogs involved selection by humans (and perhaps selection of humans by an unusual group of wolves) to optimize this type of communication. (We will not take time here to discuss house cats, who also do not recognize themselves in the mirror but yet seem to understand humans to a remarkable degree. Of course, the usual witticism is that we did not domesticate cats—they domesticated us for their own benefit.)

The wide-ranging opinions regarding the cognitive abilities of dogs might also result from the fact that the characteristics of a given dog could be greatly influenced (exactly as in the case of humans) by how that dog was brought up when it was young—something that varies widely with each dog and owner. Regardless of all these speculations, it is clear that further studies of the domestic dog (and perhaps cats) are worthwhile. For the moment, let's just say that the dog may be an exception to the proposed continuum. While they are not even self-aware, they may have a partial ToM of their human owners, perhaps because of artificial selection by humans for this very trait. It was once thought that dogs were domesticated only about thirteen thousand years ago. However, doglike fossils have recently been discovered that reportedly date back more than thirty thousand years,[21] raising the possibility that humans may have cohabited with dogs for a very long time. Some calculations based on DNA data even raise the possibility that the domestication of dogs coincided with our own origins. Outlandish as it may sound, there is a very small possibility that our canine friends actually helped us along the way to establishing full ToM. After all, if you can understand the mind of another animal and select it for its ability to understand you, you would be most effective at doing this if you yourself have a full ToM. And if you are surrounded by relatively tame wolves who can relate to you, no one else is going to mess with your tribe (or have a chance to mate with anyone in your group).

Having—I hope—satisfied dog lovers, let's return to the main question at hand. What is the real value to humans of having full ToM? At the previously mentioned Matrix of Comparative Anthropogeny website,[22] you can find a listing of human features, behaviors, and mental abilities that seem qualitatively and quantitatively different from those of our closest evolutionary cousins (see particularly the MOCA Domain for Culture, which was initially written by Don Brown, author of the classic treatise *Human Universals*).[23] If you look through this website, you will find many examples of complex human activities, behaviors, and mental abilities that should *not necessarily* require a full ToM. Examples include such things as lasting friendships, cooperative breeding, basic tool making, the construction of shelters, cooking, use of a home base, a sense of fairness, assistance with childbirth, peacemaking, sharing of food, imitation, group hunting, and so on. Of course many of these activities would be facilitated if full ToM ability were available. But having a full ToM is an *essential* ingredient for many other uniquely human mental abilities that we take for granted. The following examples are randomly chosen and presented in alphabetical order (a much longer list is obviously possible).

1. Acting (dramatics). Acting does not even require language, as an actor can just mime. But in order to be successful, actors must be aware that they are putting themselves in the mental shoes of someone else—the person they are pretending to be. The actor also has to be aware that audience members have independent minds that are capable of appreciating the performance. This is true even when the actor is in front of a camera, knowing that someone, somewhere, could either be watching live or even viewing a tape of the performance after the fact. Comedic performances are a variant of acting. Chimpanzees can manifest a form of laughter and may show amusement at physically unusual events. But for a stand-up comedian to make others appreciate humor and laugh in response, she or he has to be aware that the people seeing the performance have minds capable of understanding the humor. Conversely, those observing the performance must understand that the comedian has a mind that is independent. Note that, like acting, comedic performance does not necessarily require spoken language.[24]

2. Blushing. Charles Darwin commented that blushing might be a uniquely human behavior. It remains possible that some form of blushing occurs in other species and we just cannot detect it. But according to the standard definition, blushing requires that one be suddenly ashamed or embarrassed in the presence of others. And it is not possible to feel such emotions unless one is aware that others have minds like one's own and are therefore capable of recognizing one's embarrassment or shame. Of course it is unclear what the adaptive value is of betraying a negative self-evaluation. Perhaps this is related to the importance of sociability and accountability in human societies.

3. Care of the infirm and elderly. Humans are unique among living animals in that we take active and extensive care of the sick and elderly within our groups. In order to do so optimally, the humans doing the caring need to be aware that those being helped have minds of their own. Indeed, this can be a reciprocal behavior. Interestingly, there is evidence that some elderly Neandertals with healed severe injuries survived within their groups, indicating that our closest extinct evolutionary cousins likely had a degree of empathy and sufficient ToM to recognize the needs of an old or disabled person.

4. Concern for posthumous reputation. Many social animals pay close attention to their status, rank, and reputation within their group. But some humans also seem to be concerned about their status and reputation after they are dead and gone—their "legacy." Why would that be, since we are no longer there to witness what others have to say about us after we are dead? It appears that we can project ourselves into that distant future and imagine how others who are still alive are thinking about us—and even imagine that our own minds are there, reacting to what others are saying about us. This kind of "mental time travel"[25] is likely unique to humans, and would be hard to imagine if we did not have a ToM of others (a more realistic view of immortality comes from Woody Allen, who said: "I don't want to live on in the hearts of my countrymen; I want to live on in my apartment"). This concern for posthumous reputation might also be partly explained by the discovery of the "end history illusion"—when studied, people of all ages "believed they had changed a lot in the past but would change relatively little in the future. People seem to regard the present as a watershed moment at which they have finally become the person they will be for the rest of their lives" (and presumably beyond their deaths).[26]

5. Death rituals. All human societies carry out some form of death-related ritual, indicating that humans not only recognize the deaths of others as individuals but also recognize the fact that the person who died is no longer among us. Some other animals, such as elephants, crows, and chimpanzees, seem to have a rudimentary understanding of the death of one of their own (more on this later). But these behaviors pale in comparison to the human grief response, which stretches on from vocal private and public grieving and funerary rites to memorials, death anniversaries, and the like. These behaviors would not occur unless the individuals carrying out the rituals are fully aware that someone with a mind like theirs is no longer alive. Having developed a full ToM of other humans, we can then ascribe personhood to pets (and even to inanimate objects) and thus grieve for their loss.

6. Food preparation for others (cuisine). Although other animals do process foods, and some may even do so for the benefit of other individuals, humans are unusual in the extent to which we modify, process, and cook foods in special ways, often with the goal of pleasing others. To do this, the chef needs to be aware that the individuals for whom the food is being prepared have minds of their own. The chef also must want them to appreciate the cuisine. Note that cuisine is not the same as mere cooking, which may go further back in our ancestry—perhaps more than a million years, as suggested by Richard Wrangham.[27]

7. Grandmothering. All human societies have grandmothers. Even in societies where the average life span

is very short, there are typically a few old women surviving. Such postmenopausal women who are no longer
at reproductive age have been a puzzle from the evolutionary perspective, as the ability to reproduce is the
usual driver of natural selection. The "grandmother hypothesis" put forward by Kristen Hawkes and others[28]
is currently the best explanation for this phenomenon.
The idea is that by living past menopause a grandmother continues to contribute to the propagation of
her genes by aiding the survival of grandchildren who
are still helpless enough to benefit from her assistance.
Of course we don't know whether living longer came
first, thus exposing menopause, or whether grandmothers were first selected for living longer because of the
value of having menopause.[29] Regardless of whether the
theory stands the test of time, the fact remains that with
the possible exception of certain long-lived social female whales that have postmenopausal life spans,[30] no
other species seems to have true grandmothers. There
are certainly elephant packs and primate troops that
feature an elderly matriarch surrounded by her daugh-
ters (or possibly her sons, in the case of bonobos). But
there is so far no evidence that these old, intelligent,
highly social females specifically recognize their own
grandchildren and give selective attention to them. The
difference in humans is, of course, striking, as grandmothers typically dote upon their grandchildren even
more than mothers do and go out of their way to provide
help. Think about this in the context of ToM: It is
instinctively easy for a mother to recognize her own
offspring because of the bonding experiences of childbirth, infant care, and breast-feeding. But in order to

recognize a grandchild, the grandmother not only has to define her progeny but also know who *their* offspring are, as distinct from other young ones in the group. It seems reasonable to suggest that a full ToM would make it much easier for a grandmother to recognize her grandchildren as individuals with unique minds that need to be doted upon.

8. Healing of the sick. All human societies make attempts to heal the sick. Such attempts—although variably successful—indicate that humans are aware of the suffering of a sick person and that the sick person is an individual with a mind. The noble and underappreciated profession of nursing best exemplifies this human trait.

9. Hospitality. Almost all human societies show some form of hospitality, at least for members of their own society (something that other animals rarely do). In order to be truly hospitable, one has to recognize that those at the receiving end have minds of their own and can appreciate the hospitality. And, conversely, those giving the hospitality can expect reciprocity in the future.

10. Inheritance rules. Most human societies have inheritance rules, ranging from highly rigid to fairly flexible. To decide upon a beneficiary, you have to know that the chosen heir is a person with a mind and that you will likely die prior to that individual. Again, it is hard to imagine how this is possible without a full ToM. And in the case of those humans who leave their inheritance to a favorite pet, there was human recognition of the pet as an individual.

11. Laws and justice. The existence of specific laws and concepts of justice are common to most human cultures. Even the most tolerant societies have some defined form of punishment for inappropriate behavior. In order to develop laws and a sense of justice, people have to first know that other individuals have minds and are capable of intentional wrongdoing. This allows the society to think ahead to that possibility, generate rules and regulations, and then mete out justice when an individual fails to follow the law. Moreover, modern laws acknowledge that someone can commit a crime yet be "not guilty by virtue of mental disability" (often, such a defendant does not possess a healthy ToM). It is interesting that we humans care so much about justice that we are even willing to pay a personal price (such as risk of retribution) in order to ensure that wrongdoers get punished.

12. Lecturing. This is something that happens in all human societies (if one includes storytelling). In order to lecture or tell a story, one has to know that the audience consists of individuals with minds like one's own, even if there are perhaps minor individual differences among them. Success also requires that audience members recognize that the individual doing the lecturing or storytelling has a mind. And one of the rules of thumb for good lecturing is to put oneself in the minds of audience members. Obviously, this is difficult to do without a full ToM.

13. Multi-instrumental music. Group playing of musical instruments is common to almost all human societies. Whether in a quartet or a huge orchestra, though, participating musicians must know that the other musicians have minds of their own. This type of interaction reaches its pinnacle in Indian classical music and in the jam sessions of jazz, in which it really seems that the musicians are reading each other's minds while improvising. Of course, when there is an orchestra conductor involved, she or he is not only reading the minds of all the musicians but also expecting them to read his or her mind. Group singing in multipart harmony is a variation on the same theme.

14. Organized sports. Group sporting activities of various kinds are also found in most human societies. While play involving multiple individuals does occur in other species and sometimes follows certain basic principles (e.g., lion cubs don't actually hurt each other when roughhousing), highly organized sports with rigid and arbitrarily specified sets of rules are uniquely human. It would be hard to play such games if one does not know that one's team members also understood the arbitrary rules. In addition, trying to "read" the minds of one's opponents can be helpful. Formal sports often have a referee, and the players must understand that the referee has a mind of his or her own so that they can follow what that individual dictates. Cooperation and competition in service of a nonessential goal is the underlying notion of most sports.

15. Religiosity. Although not all humans are religious, some form of religion is universal in almost all societies (though not in all individuals in all societies). The numerous gods described by various cultures all require that devotees have ToM, as one needs to imagine the minds of gods in order to even have such a concept. The experience of having ToM could also allow humans to attribute mental states to inanimate objects, such as rocks and trees, forming the basis for animistic religions. There are of course many other cogent arguments for the adaptive value of religiosity in human evolution[31] and equally robust claims that established religions are now causing more harm than good.[32]

16. Romantic infatuation. Technically called limerence,[33] romantic infatuation is common to many cultures and ranges from special interest in another person to all-consuming love affairs. While animals can certainly become fast friends (strange combinations of carnivores and herbivores can even become friends under unusual circumstances), the widespread phenomenon of intense romantic infatuation in humans requires something approaching ToM.

17. Social control of paternity. This practice remains common in many human societies, and is often dictated by the elders in a given group. In order to control mating in this manner, the elders have to understand that younger individuals have minds of their own that need to be controlled. And when romantic infatuation conflicts with culturally prescribed mating rules, it can generate some of the great tragic love stories we are all familiar with in our respective cultures.

18. Teaching. Imitation is common in many animals, and can be taken to quite remarkable extents in some birds, such as parrots interacting with humans. But while cats may show their kittens how to play with dead mice, or chimpanzee mothers may allow their young ones to try cracking nuts, only humans take teaching to the extreme of carefully monitoring, correcting, and demonstrating specific behaviors and skills—usually to younger individuals using tested methods that vary with each society. Such active and directed instruction obviously requires the teacher to understand that the student has a mind capable of understanding what is being taught, and this knowledge must be reciprocal. Teaching also often involves slowing and exaggerating specific motions. In this case, both parties need to understand that the exaggerated motion is not what is done when the skill is actually mastered; it is used only in service of learning that mastery.

19. Torture. This horrible practice is common to many human societies and varies in its frequency, depending on social conditions and other factors. Although cats or chimpanzees might play with half-dead prey, active torture that is directed, continuous, and designed to break the spirit of the other individual (even while keeping the person alive) is a uniquely human phenomenon. Obviously, the most effective torturer is the one who can put himself in the mind of the person being tortured.

20. Trade. As Matt Ridley has emphasized in *The Rational Optimist*,[34] bartering activities of various kinds are a quintessentially human phenomenon. In order to trade effectively, one has to understand the mind of the individual with whom one is trading and to realize the pros and cons of mentally interacting with that individual—and perhaps indirectly trading with the many others who later trade with that individual.

While one may quibble with the specifics of some of these randomly chosen examples, it is clear that a great many uniquely human activities and behaviors require going beyond self-awareness and rudimentary ToM and function best with a full ToM. And many such activities also reflect the unusual human capacity for shared attention and cooperation. Indeed, Michael Tomasello has pointed out that even before they reach an age where they have full ToM, human children show a strong tendency to cooperate (something not seen in chimpanzees). Andrew Meltzoff, Patricia Kuhl, and others have documented that our apparently unique capacity for shared attention also emerges quite early in childhood.[35] While nonhuman animals can cooperate, human cooperation involves a mutual recognition of rights and duties, generated by realizing that one has a role in a joint enterprise and that others can even make a claim on one's efforts when they have played their part. In this regard, Martin Nowak outlines five mechanisms of cooperation found in nature, and concludes that humans are *supercooperators*,[36] deploying all these mechanisms to the maximum extent. Of these mechanisms, indirect reciprocity (in which I help you and therefore somebody else helps me) is particularly optimized when all parties have a full ToM, personal names, and the ability to communicate via language. The same is true of one of Nowak's other proposed mechanisms—multilevel selection, or so-called group selection (espoused with Edward O. Wilson, and currently more controversial). Regardless of the specifics, all these mechanisms are involved in various ways in what Wilson describes as the social conquest of the earth[37] by the human species. It seems that full ToM was a basic requirement for us to take over the planet.

You have probably been wondering how all this is related to the concept of consciousness, which, after all, entails being aware of your identity and place in the world. But the term *consciousness* actually has many meanings. For example, Derek Denton has suggested[38] that the earliest glimmerings of consciousness arose from simple awareness of bodily needs, such as thirst, hunger for salt, etc. Thus what we call consciousness is actually a continuum, ranging from a minimal awareness of bodily needs all the way to the full-blown mental imagery and sophistication of human consciousness. Much creative thinking on the subject of consciousness has been done by Nobel laureates Francis Crick[39] and Gerald Edelman[40] as well as many others—for examples, see books by Merlin Donald,[41] Daniel Dennett,[42] Nicholas Humphrey,[43] Antonio Damasio,[44] Christof Koch,[45] and Thomas Metzinger.[46] However, consciousness is hard to define and sometimes takes on spiritual meanings among some people, meanings that go beyond what we are talking about in this book. So we will not use the word *consciousness* going forward. We will instead stay with the terms *self-awareness* and *ToM*, keeping in mind that these states involve going through various stages in a continuum.

It would be inappropriate to write a book about the human mind without mentioning mirror neurons. These premotor neurons, discovered by Giacomo Rizzolatti and his colleagues,[47] are capable of responding both during an action and during the observation of the same action as it is performed by another individual. There appear to be sensory mirror neurons as well, which react not only when one is touched but also when someone else is being touched, suggesting that this is the basis of human empathy.[48] V. S. Ramachandran has therefore dubbed these neurons "Gandhi neurons."[49] A natural extension of this discovery is to assume that there exist neurons (or connected groups of neurons) that can both "see" and "understand" the same phenomenon. Although it is likely an oversimplification, this line of logic has led to many interesting theories about the role of mirror neurons in everything from evolution of language to autism spectrum disorders. For example, Michael Arbib[50] has written about language emergence in relation to mirror neurons, suggesting a step-by-step progression in primate evolution—from manual gesture to complex imitation of manual skills to pantomime to protosigning and finally to protospeech. The theory can also explain why humans are just as capable of learning sign languages as they are of learning to speak. I could go further and speculate that while most other primates (and other smart creatures) have mirror neurons, humans may have *mirrored mirror neurons*; that is, identical *pairs* of mirror neurons that can detect, or "see," each other. Perhaps this can help explain multiple levels of intentionality. Given the recent evidence for preferential excitatory connections among groups of sister neurons, one even can posit the existence of sister groups of mirrored mirror neurons in humans.[51]

Regardless of the specifics, mirror neurons (or a neuronal mirror system of some kind) could certainly end up being a major part of the explanation for the phenomenon of self-awareness and perhaps the theory of mind. But this book will not go much further into this subject, partly because the amount of reliable data about human mirror neurons is limited.

In this regard, let's conclude this chapter by considering a very common variation from the "typical" human mental state—a variation in which ToM and intentionality seem to be altered, leaving the subjects unable to effectively carry out many of the twenty typical human behaviors listed above (or others like it). We have earlier mentioned autism,[52] a mental difficulty affecting children and young adults that some people claim is increasing in frequency and severity in Westernized societies. Whether or not the increase is real (a subject of some controversy), the word *autism* has been largely replaced by the term *autism spectrum disorders* (ASD), for the reason that autism is not a single disease. Rather, like "pneumonia" (any kind of disease that fills up the lungs and makes you short of breath), ASDs are a group of diseases that can be lumped together because they have certain common features. The spectrum ranges from individuals who are completely unable to communicate and take care of themselves all the way to so-called high-functioning autistics who are only slightly "abnormal" compared to others (a condition that is sometimes called Asperger's syndrome). But one common feature that runs through all cases of autism spectrum disorders is the relative or absolute lack of ability to communicate with others at a social level (a social communication deficit). In fact, this is often the first feature noticed by parents, and Leo Kanner's description of classical autism in 1943 described individuals who had an "inability to relate themselves in the ordinary way to people and situations from the beginning of life." In other words, autistic subjects partially or completely lack the ability to imagine themselves in the situations of others (sometimes called mind-blindness). Effectively, then, autism spectrum disorders include a difficulty with full "attribution of mental states to others." In other words, many of these individuals appear to be lacking a full ToM.[53]

In the latest update of the *Diagnostic and Statistical Manual of Mental Disorders*, published by the American Psychiatric Association (DSM-5, scheduled for release in 2013), two cardinal features are being defined as typical of ASD:[54]

1. persistent deficits in social communication and social interaction across contexts, not accounted for by general developmental delays and manifested by all three of the following: deficits in social-emotional reciprocity; deficits in nonverbal communicative behaviors used for social interaction; and deficits in developing and maintaining relationships appropriate to developmental level; and

2. restricted, repetitive patterns of behavior, interests, or activities as manifested by at least two of the following: stereotyped or repetitive speech, movements, or use of objects; excessive adherence to routines; ritualized patterns of verbal or nonverbal behavior or excessive resistance to change; highly restricted, fixated interests that are abnormal in intensity or focus; or abnormal reactivity to sensory input or an unusual interest in sensory aspects of the environment.

To complete the diagnosis, the symptoms must be present in early childhood (but need not be fully manifested until social demands exceed capacities) and must limit and impair everyday functioning.

As you can see, one cardinal criterion is a social communication deficit (effectively, varying degrees of deficits of

ToM). In this regard, Simon Baron-Cohen confirmed to me that blushing is anecdotally absent in individuals with autism (but that this needs further study). In a related domain, such individuals also perform more poorly than expected on tests of recognition of faux pas, a recognition that normally entails feeling embarrassed for someone else.[55] Baron-Cohen also noted that parents may complain about teenage children with autism undressing in public without embarrassment, something one would never see in a typical child.

The second cardinal criterion is repetitive behaviors that often manifest themselves as unusual talents. A small percentage of individuals who behave this way have savant syndrome, as we discussed earlier—a condition in which remarkable abilities are concentrated in a few narrow domains of human cognition. Of course, there is a tendency to equate the independent existence of unusual talents with being "on the spectrum," leading to suggestions that many historically famous scientists such as Newton and Einstein were actually autistic. In fact, being "on the spectrum" is even considered to be a positive attribute in places like Silicon Valley, where employers may want to hire bright individuals who prefer to spend most of their time in front of a computer screen rather than socializing with others. This is scant comfort to the parents of the unfortunate children who have severe communication deficits but do not appear to have any obvious special talents. Of course, as Francesca Happé and Uta Frith so eloquently put it in their phrase "the beautiful otherness of autism,"[56] these minds are just *different* from ours, and we should probably be making a greater effort to find and foster such hidden talents.[57] One example of this is Jake Barnett, who was featured on the popular TV show

60 Minutes. Jake was diagnosed with autism at age two and became increasingly withdrawn and uninterested in most things. All attempts at conventional therapy failed until his mother happened to take him along to some evening classes involving mathematics, physics, and astronomy. He then began to come out of his shell, asking questions and learning so avidly that he was eventually able to graduate from high school and enter college at a very young age. Jake appears destined for brilliance in mathematics and physics. Interestingly, the process by which he came out of his shell in the setting of mathematics and physics was associated with his also recovering from his most severe autistic symptoms. Perhaps there are many more such hidden talents that have yet to be uncovered among cases of autism. Indeed, one father of an autistic child has created the Specialisterne (which translates from Danish as "the specialists")—an unusual social enterprise, staffed mostly by people with autism, that provides assessment, training, education, and IT consultancy services. The Specialist People Foundation, which runs Specialisterne, says it has a goal of enabling one million jobs for people with autism and similar challenges through social entrepreneurship, corporate-sector engagement, and a global change in mind-set.[58]

Another remarkable feature of ASDs are their consistent and widespread occurrence in human populations worldwide, despite the fact that they are poor prescriptions for reproductive fitness. As most severe autistics do not leave progeny, why was the condition not eliminated by natural selection? The current explanation is that most causes of autism represent de novo (new) genetic mutations, which can then be inherited in a family.[59] This begs the question of why such mutations should be happening at such high frequencies in otherwise critical genes. One possible explanation is that parents are having children at an older age than they did in previous generations—especially fathers, whose sperm are at greater risk for accumulating random mutations over time.[60] Another possibility is that the concentration of high achievers into certain geographic areas and professions (such as the tech sector) may be increasing the possibility of individuals already "on the spectrum" mating with each other.[61] In addition to such modern possibilities, our evolutionary transition to a full ToM may be far from completely hardwired (since it requires an environmental input at the right time), and the risk of failure of the ToM mechanism remains very high in the human species. Yet another view is a more positive one, and posits that having group members who were able to ignore social niceties and instead deal with specific problems and difficult tasks (in other words, having members who displayed the features of mild to moderate autism) may have actually been of benefit to early human societies. As Aristotle recognized, humans are the greatest imitators. So once someone in a group is able to invent or discover something, imitation will rapidly spread the new concept through human populations. Thus having a few individuals with less-than-optimal ToM could benefit the group, since their ability to perform restrictive and focused activities would eventually be of value to others. And such individuals might also be the ones who could best memorize and accurately pass down important details of a lengthy oral tradition before the advent of writing. If this scenario is correct, it may help explain the persistence of genes predisposing people to ASDs in all societies and cultures. Severe autism spectrum disorders may well be the price that we pay for the benefits of having milder cases. In keeping with this, Temple Grandin says that if not for the talents of many people with various levels of autism over the years, we humans might still be living in caves!

Of course, such considerations of autism remain limited in discussions of human origins, for obvious reasons. First, one cannot assume that such individuals represent the mental state of humans prior to the final transition to modern cognitive behavior. Second, many of these cases feature other neurological, psychological, or cognitive abnormalities, making it difficult to come to any easy conclusions. Regardless, the social-cognition deficits of autism seem to involve a difficulty with expressing a full ToM. Thus it is reasonable to suggest that a better understanding of autism would also lead to a better understanding of ToM. Recent studies have made it clear that inherited or de novo gene defects, duplications, and deletions play a key role in the causes of autism and that any single genetic cause accounts for less than 1 percent of all cases. It appears that many different defects can trigger the neural processes involved in generating the symptoms of autism. There are many hypotheses regarding these processes, ranging from the "extreme male brain" theory (Baron-Cohen)[62] to abnormalities in nerve cell number or pruning (Courchesne and others)[63] to defective mirror neurons (Ramachandran) to the social motivation theory (Chevallier and others)[64] to autism as the diametric opposite of schizophrenia (Crespi).[65] To adequately address these and other theories would be far beyond the scope of this book.

Regardless of which of the theories turn out to be right (and it is very likely that more than one will be), social communication deficits can be equated partly with variable degrees of lack of ToM—a feature we have suggested was a critical step in the evolution of the human mind. Thus, in a sense, some individuals with autism spectrum disorders might be studied to test our primary hypothesis—that the evolution and stabilization of full ToM in human ancestors was originally held back by the recognition of mortality of others, which in turn results in awareness of personal mortality (a distressing and maladaptive situation). If so, one might predict that individuals with severe autism spectrum disorders (i.e., those with very limited ToM) should not be fully aware of the meaning of death. As you can imagine, this is not an easy question to ask or address, and there seem to be no definitive published studies on the matter. But anecdotal evidence suggests that autism spectrum disorders are indeed associated with a relative lack of full understanding of the death of others. One can even find Web-based instructions for parents on how to explain the death of family members to individuals with ASD. There are also no formal studies that ask whether autistic individuals understand their own mortality. Again, it would be difficult to get such a study approved—to directly ask such individuals whether they have a true recognition of their personal mortality. Regardless, it is interesting to ask how many of the twenty random examples of uniquely human features requiring full ToM we listed above are significantly blunted in autism spectrum disorders. It would likely be most of them.

Given how useful and powerful full ToM has been for the success of our species, why did some of the other intelligent and self-aware species not acquire the same ability? Was it just chance? Or a very special one-of-a-kind brain evolution process that was specific to the human species? Or is there simply a nearly insurmountable "wall" that has been holding back all the other species? If so, what is that wall?

CHAPTER 5

The Wall

He had no conscious knowledge of death, but like every animal of the wild, he possessed the instinct of death. To him it stood as the greatest of hurts. It was the very essence of the unknown; it was the sum of the terrors of the unknown, the one culminating and unthinkable catastrophe that could happen to him, about which he knew nothing and about which he feared everything.

—Jack London, in *White Fang*

The human race is the only one that knows it must die, and it knows this only through its experience.

—Voltaire

As mentioned earlier, anatomically modern humans with skeletons like ours began roaming Africa around two hundred thousand years ago. But the critical transition in our mental evolution toward attaining full ToM may have occurred more recently. So far, archaeological evidence of humans with full ToM (such as representational art, personal ornaments, and artifacts associated with ceremonial burial of the dead and trade) is about one hundred thousand years old, and the earliest evidence is found in Africa or in nearby sites. This is why the term *behaviorally modern humans* is sometimes used to denote the emergence of our own species, which eventually replaced all other humanlike species as we spread across the planet. It just so happens that genetic evidence indicates that our own species—which began with an effective population size of only about five to ten thousand individuals, also in Africa—diverged from other humanlike species about one hundred thousand years ago as well.[1] As we discussed at the end of the last chapter, many features of modern human behavior likely required a full ToM. After all, why would you make a bead necklace for yourself unless you knew that others had minds like yours to appreciate your self-decoration? Or why place funerary items into a burial site unless you recognized that the dead individual was once another person with a mind like yours? We suggest that during the time from around one hundred to two hundred thousand years ago, our anatomically modern ancestors encountered the same barrier that had effectively prevented the further mental development of all other intelligent animals. This wall is also hit whenever a *single* individual develops a full ToM and becomes fully able to attribute mental states to others. This chapter describes why this situation presents a difficulty for natural selection to surmount. But before moving to this issue, it is useful to summarize some salient points from the previous chapters to ensure that all readers are on the same page.

1. A fundamental tenet of evolution is that natural selection will act to maximize the fitness of individuals in a population, where fitness is defined as the ability

to reproduce successfully and pass genes on to the next generation of the population.

2. The basic vertebrate body plan, including the overall organization of our brains, was established hundreds of millions of years ago.

3. There is ample genetic variation in large populations—enough to enable more elaborate versions of structures to evolve by natural selection if there is sufficient positive selective value in doing so. That is, the genetic parts necessary to make greatly exaggerated versions of our limbs, senses, organs, etc., are already there.

4. There should be nothing special about brains with regard to the general principles of natural selection and evolution.

5. Many animals are quite intelligent, especially long-lived individuals living in groups with hierarchical social organizations. In such animals, intelligence should generally be a good thing, and more intelligence should be even better.

6. Being smarter will be beneficial in many ways, but, like all exaggerated traits, too much intelligence will come with some costs, such as mental confusion and obsession with details. But evolution has dealt with such costs by developing higher brain functions that permit conceptualization and generalization. Again, these should not have presented any unique obstacles for the evolving of even more intelligent species.

So from both the genetic and natural selection points of view, there is no obvious reason why it took tens of millions of years to evolve from an animal with full self-awareness and rudimentary ToM to one with humanlike intelligence and a full and extended ToM. And once such a level of intelligence was reached, a species should then become much smarter rather quickly (in evolutionary time, as seems to have happened with humans). Taken together, these observations suggest that there was a specific barrier at some point along this evolutionary road. Breaking through that barrier required an unlikely series of events—and we suggest that it happened only once on this planet, in behaviorally modern humans, that is, us.

Once the barrier is breached and full ToM is established, a species can continue down the path to greater intelligence essentially unrestrained, becoming much smarter than those still held in check—in other words, conventional natural selection would once more swing into action. But rather than assuming that emergence of humanlike cognition was a gradual, step-by-step, or piecemeal *positive* selection of a series of useful changes, we present a counterintuitive view—that the initial *negative psychological cost* to the *individual* of attaining a full ToM and a more complete appreciation of reality (especially awareness of one's mortality) was too great to allow the first such individual to get past it, reproduce, and successfully propagate that ToM ability within a species. And the situation stayed this way in many species, until humans finally broke through the barrier by simultaneously generating mechanisms for denial of reality and mortality. And once we did, huge opportunities opened up.

Let's reflect on the implications of the steps through which a full ToM might evolve. An animal has full self-awareness when it reaches a level of intelligence that drives it to recognize and examine its own identity, its place in the world relative to the rocks, trees, and so on. It then develops rudimentary ToM when it reflects on its own situation relative to its brethren. The nervous system is now able to move well beyond reflexive or calculating pathways, and the brain can progress to reflective thought, perhaps beginning to question what it means to be alive. At this stage, an animal with rudimentary ToM can recognize that other individuals are intentional agents, that is, that they are capable of independent and intentional actions. But when a single member of such a species goes further to develop a full ToM, this individual will fully realize that others of its own kind have minds and personhood, and will only then begin to reflect on the meaning of their deaths. This appreciation for the implications of death has major repercussions for natural selection.

At first glance an evolutionary biologist may still see this as a strange statement. Why should being aware of one's mortality have a major impact on natural selection? After all, natural selection simply selects from existing variants for whatever functions best in the environmental and competitive situation a given population finds itself in. The idea we present is that crossing this *psychological evolutionary barrier* into maintaining a full ToM would have inevitably resulted in selection *against* such individuals—because the simultaneous awareness of mortality would result in such individuals being rapidly eliminated from the population. Why should this be so? Such an animal would already have built-in reflex mechanisms for fear responses to dangerous or life-threatening situations. But this unconscious fear would now become a conscious one, a constant terror of knowing one is going to die, and that it could happen anytime, anywhere. In this model, selection would only favor the individual who attains full ToM *at about the same time* as also achieving the ability to deny his or her mortality. This combination would be a very rare event. On top of that, more than one individual with this combination would likely have to be born at about the same time and then find and mate with the other to have a chance of genetically establishing these new abilities in a cohort of their progeny ("fixing the genetic basis for the abilities in the population" would be the technical term). It is even possible that this was the defining moment for the original speciation of behaviorally modern humans. Again, the novel idea is *negative* selection *against* achieving full ToM, as opposed to conventional *positive* selection for this trait. In other words, unless one could first get past the immediate negative consequences of a full ToM (particularly awareness of one's own mortality), the likelihood of ever getting to use the positive features of ToM are slim to none. This is the Rubicon that we humans seem to have crossed over.

Just as with the animals who first made the difficult transition from living in water to living on land, many other preexisting conditions had to be just right for this psychological crossing to happen. Let's consider some of these conditions. Social selection is a general concept that refers to the social cognition, social brain, and runaway social selection theories espoused by various authors, which suggest that ongoing complex social interactions among mentally sophisticated animals would have increased the likelihood of their developing increased intelligence. This was clearly a critical step along the way. But this argument should apply to any intelligent social animal, of which there are many. Thus while such mechanisms are a key part of the human origins story, they are likely to have preceded the psychological evolutionary barrier that we have focused on. The same is true of Sarah Hrdy's related theory,[2] that helpless human babies can only survive the dangers of infancy by establishing relationships with their mothers and with an extended group of caregivers, all of whom also need to relate to and understand each other. This could indeed have been an earlier mechanism for the propensity toward having full ToM, and could have worked in combination with social selection. A complementary feature would have been empathy and basic altruism,[3] which could have been initially driven by the maternal instinct (arguably, the evolutionarily earliest and strongest form of empathy). This in turn could have been aided and abetted by the role of the hormone oxytocin, which encourages such affiliative behaviors (see Patricia Churchland's *Braintrust* for a discussion of the potential role of oxytocin in altruism).[4] While all these features would only require rudimentary ToM, they would *also be greatly enhanced by attaining a full ToM*. The same is true of the strong tendency of humans to cooperate, and of our apparently unique capacity for shared attention.

All these factors and more would have worked together to set the stage for the benefits of full ToM. But each time one individual among our ancestors managed to achieve a full stable ToM, she or he would have faced the psychological barrier of watching others die and thus becoming aware of his or her own mortality.[5] A reasonable mechanism for eventually crossing this barrier is the ability to hold a false belief about a situation, that is, *denying reality*. Indeed, as prominently discussed in Robert Trivers's *The Folly of Fools*, deceiving others is best done if one can first deceive oneself.[6] *The Folly of Fools* is a follow-up to Trivers's classic work of many years ago[7] and brings up to date the evidence that deception occurs at all levels in biology and plays a vital role in many evo-lutionary interactions. As Trivers points out, deception and self-deception took a special leap forward in humans because of our abilities for language and communication. Less defined in his latest treatise is the critical role of full ToM in these abilities. Deceiving others would be much easier if we knew that those we are deceiving have minds not too different from our own. And the ultimate form of reality denial is the im-plantation of a false belief in one's own mind.

Exactly how did this denial of reality come about? We will discuss this later, but while other mechanisms are possible, it seems most plausible that the process involved partial loss of an existing neural mechanism rather than the addition of a new one. As a consequence, other forms of reality denial became part and parcel of the human condition. And, as we shall discuss later, the ability for deception and self-decep-tion could also have been important for reproductive success via improved mate choice.

You are probably still thinking that the proposed barrier should not be *that* difficult to surmount. After all, shouldn't normal humans be able to rationalize their way through this problem by using their normal thinking skills? But you really have to consider the situation *not as it is today* but *as it was at the time when the first human ancestors were trying to cross the barrier*. Perhaps this distinction will be more ev-ident if you consider another self-aware species. Imagine a particular elephant in Africa today, which via some unique genetically based neural changes has achieved full under-standing of the minds of other elephants (full ToM). At first glance this would seem to endow the elephant with very use-ful new powers, making it better able to understand, control, and even deceive others. But this individual would not immediately have any other individuals with similar abilities to interact with, and might find the situation a bit unnerving. This disturbing feeling would be much aggravated when another elephant dies and the one with full ToM comes to realize what this means—that she or he is also going to die. This realization of mortality would then trigger built-in mammalian fear responses, generating a disabling psychological terror that is not easily overcome. This elephant would now behave strangely and would be treated differently by other members of the group. Coupled with a disturbing fear of mortality, such an individual would be unlikely to engage in or succeed in competition for opportunities to mate and reproduce (such as the dangerous physical combat needed for male mating success). It would therefore be unlikely to pass on its genes to the next generation—an evolutionary dead end. In other words, individual *personal survival* would take precedence over typical behavioral drives that ensure *species survival*. Such evolutionary dead-end episodes may have been occurring off and on for many millions of years in individuals from various species who already had basic self-awareness and rudimentary ToM but then failed to pass on their newfound ability for full ToM to their progeny. It is, of course, difficult to prove this, and we may never know for sure. Regardless, humans as a species clearly achieved full ToM and then spread across the planet, taking with them this ability wherever they went. Exactly when, where, and how our ancestors originally achieved this transition is discussed later.

Having now stated the theory in detail, one must ask if it is really brand new. As always, it turns out that apparently original ideas are either not unique or have their roots in prior human thoughts. The observation that the fear of mortality is not a serious psychological difficulty for humans actually dates back thousands of years. For example, the Mahabharata, a Vedic text of ancient India, reports that humans were expressing surprise at our ability to deny the reality of death more than four thousand years ago. In the classic myth, the five Pandava brothers are out on a hunt, become desperately thirsty, and one of them goes ahead in search of water. Sahadeva finds a beautiful lake but encounters a yaksha (spirit), who warns him not to drink the water until he answers some questions. Sahadeva ignores the yaksha, drinks the water, and drops dead. A similar fate befalls three other brothers, one by one. The eldest brother, Yudhisthira, finally comes by and demands that the yaksha restore his brothers to life. The yaksha again requests answers to his questions, and Yudhisthira agrees. The yaksha then asks many questions, spanning a wide spectrum of practical and philosophical issues. Yudhisthira answers them successfully, and the yaksha restores his brothers to life. Toward the end of the series of questions comes the one of relevance to us.

> The yaksha asked: "What is the greatest surprise?"
>
> Yudhisthira replied: "People die every day, making us aware that men are mortal. Yet we live, work, play, plan, etc., as if assuming we are immortal. What is more surprising than that?"

So the first part of the concept (that humans are surprisingly indifferent to their mortality) dates back to prehistory. Another related concept from ancient Hindu teachings concerns three situations in life in which involuntary mental detachment from the real world occurs and one begins to deeply ponder the meaning of life, often resulting in sincere resolution making (called *vairāgyam* in the Sanskrit language).[8]

One of these situations occurs when a person is present at a funeral. In this circumstance, one tends to become detached, focusing on the futility of life and the fact that one will also die someday and leave this world. In this state of detachment one may resolve to hereafter live a more decent, honest, and pious life. But upon leaving the funeral this *śmaśāna vairāgyam* is quickly over, and one forgets about death and continues to live as one did before. You may have had the same experience when attending the funeral of someone who was not particularly close to you. On the way back, thoughts of your own mortality, its implications, and any resulting resolutions (if you had them to begin with) are already fading from your mind, and you may be thinking more about what you will eat for dinner that day.

The *prasūti vairāgyam* situation occurs when a woman is giving birth, suffering intense pain, wondering why she is going through this agony, and resolving to avoid this torture in the future. But yet again the negative feelings fade away rapidly after the labor process is over. A woman friend who read a draft of this book said the pain of her first childbirth was so intense that she asked her husband to remind her how awful it was and to never again repeat the torture of going through labor. And then after her second child, she wrote herself a note so she would remember. But then she had a third—and she finally got the message after that. Indeed, as early human groups began to understand the connection between sexual activity and pregnancy, women must have evolved the ability to have this sort of amnesia about the severity of pain during childbirth. Otherwise they would have sought to avoid sexual activity after the first pregnancy—not a good prescription for reproductive fitness and the propagation of the species.

The *Purāna vairāgyam* situation occurs when people listen to inspiring speakers or teachers who talk about the impermanence of life and its meaning. As you listen to such ideas you fall into a deep *vairāgya bhāva* (feeling of detachment), ponder the meaning of your own life, and perhaps make some resolutions for the future. But once you leave the lecture, your mind quickly returns to its usual state. We hope you experience a *Purāna vairāgyam* while reading this book, but we realize that this will probably fade when you finish reading it and get back to your routine.

You may have heard of the classic Greek myth of Prometheus, who gave mankind the gift of fire. For this and other transgressions, Zeus (the king of the gods) punished Prometheus by having him chained to a rock, where an eagle came every day to tear at his liver. In Aeschylus's *Prometheus Bound*, the chorus (usually an interactive participant in Greek plays) asks Prometheus what else, in addition to fire, he has given to humans.

> Chorus: Did you perhaps go further than you have told us?
> Prometheus: Yes, I stopped mortals from foreseeing doom.
> Chorus: What cure did you discover for that sickness?
> Prometheus: I sowed in them blind hopes.

Academically oriented readers would probably like to see here a detailed exposition of the stages in human thought that took us from these prehistoric times to the present day. But we will keep this brief for the general reader and provide just a few examples. Indeed, philosophers and psychologists through the ages, including Kierkegaard and Otto Rank, have put versions of similar ideas forward. A modern variation of the concept was presented in Ernest Becker's book *The Denial of Death*, which pointed out that many otherwise inexplicable aspects of human culture and behavior can be explained by our pressing need to deny our mortality. Becker's followers include a group of psychologists (Sheldon Solomon, Jeff Greenberg, and Tom Pyszczynski) who coined the term *terror management theory*, which proposes that

> the juxtaposition of an inclination toward self-preservation with the highly developed intellectual abilities that make humans aware of their vulnerabilities and inevitable death creates the potential for paralyzing terror. One of the most important functions of cultural worldviews is to manage the terror associated with this awareness of death. This is accomplished primarily through the cultural mechanism of self-esteem, which consists of the belief that one is a valuable contributor to a meaningful universe.[9]

These authors have provided experimental evidence that humans indeed have a variety of worldview and self-esteem mechanisms to deal with the terror of knowing they are going to die (sometimes called mortality salience). In a 1996 discussion[10] of stroke patients who deny the reality of their paralysis (anosognosia), V. S. Ramachandran pointed out that "humans are unique in ... [being able to] ... contemplate and plan for the future ... but the penalty they had to pay for this is the awareness of their own mortality." Ramachandran then asked, "Why isn't everyone paralyzed by the constant fear of death?" and suggested an explanation: "Our brains also evolved a special purpose mechanism to selectively uncouple [awareness of death] from limbic/emotional centers in the brain, a mechanism akin to denial or 'repression.'" The Nobel Prize–winning physicist Richard Feynman discusses his wife's death and writes:[11] "If a Martian (who, we'll imagine, never dies except by accident) came to Earth and saw this peculiar race of creatures—these humans who live about seventy or eighty years, knowing that death is going to come—it would look to him like a terrible problem of psychology to live under those circumstances, knowing that life is only temporary. Well, we humans somehow figure out how to live despite this problem: we laugh, we joke, we live" (Feynman likely wrote this as he himself was struggling with cancer). New York City psychiatrist Robert Langs has written books[12] in which he suggests that "trauma and the threat of harm and death are the unconscious driving forces behind our most critical decisions and choices," and that death anxiety exists in three major forms, which "unconsciously motivate our most creative and most devastating actions and decisions." He goes on to state: "There is only one defense against existential death anxiety—denial, which banishes these feelings from consciousness and into the deep unconscious." British philosopher Stephen Cave calls this the mortality paradox,[13] stating that "Death therefore presents itself as both inevitable and impossible . . . it involves the end of consciousness, and we cannot consciously simulate what it is like to not be conscious." And psychologist Nicholas Humphrey suggests, "If fearing death is one of the consequences of being conscious, and if fearing death helps to keep human beings alive, this suggests that consciousness—core consciousness—contributes more to the biological fitness of human beings than to that of any other animal. It could even mean that human consciousness, at this basic core level, has in fact come under new pressure from natural selection."[14] Perhaps related to all this is the discovery by Tali Sharot and her colleagues of specific neural mechanisms that mediate unrealistic optimism in the face of harsh reality[15] (as we said earlier, what is optimism, after all, but reality denial?).

There are many other versions of such ideas in the thoughts and writings of humans over the eons. However, all these represent only one part of the theory presented here. If one accepts that it is surprising that we humans deny our mortality despite being aware of it, we must ask what the evolutionary origin and purpose of this unusual form of denial is. Some will still argue that the uniqueness of the human mind is evidence that our brains are fundamentally different from those of our animal cousins. That is, we are not just a little self-aware; we are smart enough that we have become obsessed with who we are, why we are here, and so on. So there must be some special brain mechanisms or neurons or connections that we have not yet discovered. But if a mammal or bird already has the neural circuitry to make a mental picture of its own mind, why should it be so difficult to evolve a similar circuitry for a mental picture of another similar mind?

This is a good point at which to discuss clues we can glean from studies of childhood mental development. As we mentioned earlier, we humans go through an unusually long period of development after birth, starting out as completely helpless infants and eventually maturing into functional adults after many years (the standard quip from long-suffering parents is that maturation takes about thirty years—and actually, data on prefrontal cortical brain comparisons suggests this may not be too far off the mark, since maturation seems to be continuing even into the midtwenties, at least in humans, but not in chimpanzees).[16] This protracted period of development is rather unusual among animals and likely contributes to human uniqueness by providing a long period of brain growth after birth, during which brain development and learning are still continuing. The question arises whether the mental stages through which we progress during our development after birth in any way recapitulate the stages that our ancestors went through during their evolution. At first glance it may seem that we are resurrecting the now somewhat discredited phrase "ontogeny recapitulates phylogeny." This concept came from Ernst Haeckel (1834–1919), who posited that the stages through which a vertebrate embryo develops recapitulate the evolutionary stages through which the species had evolved. While there is a germ of truth to this idea, it was overstated, and some of Haeckel's famous drawings may have been modified somewhat to suit the hypothesis. However, before throwing this metaphorical baby out with the bathwater, perhaps we should consider the possibility that human postnatal mental development (*cognitive ontogeny*) might recapitulate the recent evolution of human minds (*cognitive phylogeny*). Could it be that our ancestors went through stages of mental evolutionary development not too dissimilar to those that each of us now goes through during childhood?

Regardless of what you think of this suggestion, it is interesting to consider the stages of human mental development after birth in relation to the understanding of mortality. Although experimental literature on this subject is understandably limited (and there are major confounding effects of culture, particularly religion), a general pattern emerges. Evidently, children do not concern themselves much with death in the early years, even after they have developed self-awareness (between one and two years of age) and the first inklings of ToM emerge (between three and four years of age).[17] Even when they are between the ages of five and seven, children think of death as some kind of state of sleep,[18] and are unclear as to what exactly happens when someone they know dies (in a religious family they would have been taught to believe whatever their religion says about death). Shortly thereafter children begin to ask questions about death and what it is all about. There is also sometimes a period of concern and fear over death per se. However, this phase does not last very long, and children soon develop the same denial of mortality that adults seem to have. They then progress to the stage of "invincible teenager," an individual who is not scared of very much at all (the many dangerous and sometimes fatal pranks and games played by young people attest to this). In fact, as mentioned above, some studies suggest that the human brain is not yet fully developed in the teenage years. The amount of gray matter in the frontal lobe of the brain is changing during adolescence and even during early adulthood, and myelination (the development of insulation on the wiring of the brain) is continuing during this period—in contrast to great apes, in whom this process is complete by puberty. In keeping with this, careful psychological testing shows that teenagers do not yet have the full "perspective-taking abilities" that adults have.[19] It is difficult for a teenager to put himself or herself fully in the mind and perspective of another individual (parents of teenagers would no doubt say they already knew this). Of course, humans do gradually become more risk-averse and sensitive to mortality from middle age on. And having children often leads people to focus on avoiding risks as well.

As it happens, the stages through which we humans pass during our growth and childhood do mirror the stages we have suggested in the evolution of human understanding of mortality. Perhaps this is indeed an instance in which *postnatal cognitive ontogeny recapitulates recent cognitive phy-*

logeny. Could it be that *Homo erectus* had the self-awareness of a two-year-old, and that Neandertals and other anatomically modern humans only had the rudimentary ToM of three- and four-year-olds? And is it possible that the emergence of behaviorally modern humans (us) involved a transition across a barrier, so that we became six- and seven-year-olds with full ToM who fully understood death, and then went on to having adolescent minds with no fear of death? I discussed this scenario with Barry Bogin, an expert on human growth and development. After thinking about it further, he wrote me as follows: "I suspect that the full pattern of modern human growth and development, with adolescence and the adolescent growth spurt, is needed for the full package of human cognitive, social, economic, political, and cultural capacities. Even more speculative—perhaps the 'break-through' to a full theory of mind that you and I discussed required an adolescent stage to reorganize the body, the brain, our emotions, and our sociocultural behavior to accept and successfully deal with the combined forces of knowing we will die and denying reality." Interestingly, based on all the fossil data to date (including that of Neandertals), Bogin and Holly Smith state: "The most parsimonious conclusion that one may draw . . . is that all the features of the modern human adolescent stage of the life cycle evolved only in the *H. sapiens* line. Quite likely this would be no earlier than the appearance of archaic *H. sapiens* in Africa."[20] Overall, while it is impossible to prove that postnatal cognitive ontogeny recapitulates recent cognitive phylogeny, this happens to fit all the facts. Once the psychological barrier was successfully breached, natural selection and sexual selection could then have taken advantage of the situation to quickly evolve a new species with fully human minds.

Are you still not convinced that the isolated acquisition of full ToM would be so counterselective to species survival? Another hypothetical example may help. Assume that you are a young adult male lion. Like all the other young studs, you are pretty frustrated. That big old guy with the flowing mane has his way with all the lionesses in the pride, and if you want to get some action you have to try to sneak in occasionally without being noticed. This doesn't happen very often, if at all, and you decide you want to try to do something about it. If you want to sow your oats freely, you are going to have to challenge the dominant male. Although many animals have evolved stereotyped combat rituals that are evolved to select the "best" male while limiting danger to the combatants, there remains a significant chance that you will be killed or badly injured. What do you do? If your brain is dominated by thoughts and emotions that have been shaped by natural selection, you will most definitely go for the fight if you feel there is even a slim chance of victory. Winning the battle will mean that you can mate at will with multiple females and that your genes will be shooting into the next generation by the bucketful. Your fitness, as measured by your reproductive success, will increase tremendously.

However, if you are that very rare young lion with full ToM who has thought about your mortality at length, you will realize that if you are killed, that's it—your life will be over. Bummer! It might be nice to have more little lions like you, but it will be a personal disaster if you fail instead and cease to exist. From a rational perspective, it would be crazy to risk getting killed in order to further the existence of your genes. Like all animals, you would already have an innate form of death-risk fear, a trait that would have been positively selected for in past generations of your species. So after becoming aware of mortality, it would seem even more rational to avoid death risk, thereby increasing the chances of personal survival. And going back to our discussion on the dangers of the human labor process, it would be easy to imagine the reasoning of a female with ToM who desires sex but fears death in childbirth. Without denial, the woman might well choose to avoid sex rather than take the risk.

This is the fundamental problem with full ToM and the corresponding recognition of mortality. Life places us in situations where fitness is at odds with survival. If an animal thinks about what it really means to be dead, the rational one, who understands this reality, will choose survival. But natural selection chooses reproductive fitness over personal risk. Thus that little extra level of smarts that brings on understanding of reality and mortality comes at a great selective disadvantage in the population. Members of the group who still think little of mortality will survive and reproduce better. In this argument, those individuals who first become fully able to attribute mental states to others (achieve full ToM) and thus become aware of their mortality are at greater risk of losing out, because they *avoid mortality risk over opportunities to reproduce* and therefore will not pass their genes on as effectively. Of course, survival is a component of fitness—animals have to live in order to reproduce. This is especially true before we reach sexual maturity. However, life is eventually going to place creatures in situations where their survival will be placed in some jeopardy if they pursue the course that will maximize their ability to reproduce. Indeed, in many animals reproduction itself is risky. The resources devoted to growing offspring, both within and outside the womb, are draining for parents, and childbirth itself carries significant dangers, especially in humans.

The dilemma of the lion with ToM is one in which reproduction is directly tied to a risky adventure. But, unlike other animals, humans are also willing to risk death every day for seemingly trivial reasons. We drive our cars too fast, often refusing to use such simple protective measures as seat belts. We go to war, where we can expect to have others aiming to kill us. The newspapers are full of stories of persons who engage in lethal confrontations over trivial issues. Natural selection goes against all these decisions to some degree, based on a biological "calculation" that compares the value of survival and fitness in each case, but the calculation is not being done via any conscious, rational reasoning. The behaviors of animals are determined by unconscious biological calculations, but when an animal attains full ToM aided by reality denial, a whole new dimension is added to the equation.

We previously mentioned that emotions can also short-circuit rational analysis in humans and other animals. Could we not have just extended our emotional repertoire gradually to deal with emerging ToM, similar to the way in which—as we proposed earlier—our executive functions gradually developed to compensate for increasing mental ability? The problem with this is that some emotions may be cranked up sufficiently to make us do things that are not in the best interest of survival, such as the kind of love that might make a mother devote most of her resources to raising offspring. Moreover, many emotions are already working largely to promote survival, and the struggle between irrational drives (e.g., fear and lust) is often the calculator that determines whether we will optimize fitness or survival.

Our emotions are evolved to produce a quick, generally useful response to a circumstance; to maximize our ability to reach a goal. Full ToM does not simply introduce an incremental change in how we should weight various inputs but rather turns the entire system on its head. It suddenly changes the ultimate goal that our behavior is trying to attain. The entire system of behavioral drivers in an animal on the verge of full ToM has evolved to generate behaviors with a single end point—maximal reproductive fitness—which translates at the population level to *species survival*. Full ToM creates a very different new target of *personal survival*. And some of our most influential emotions, such as fear, are extremely capable of directing our behaviors to move us toward that goal. Our built-in fear response will be instantly heightened to enormous levels if we are suddenly obsessed with the consequences of death. What emotion or unconscious behavioral driver can compensate for that fear—or, more accurately, could possibly have already been in place to compensate for that fear—when our ancestors first attained ToM?

Loss of risk aversion should have gone through two phases. In the first phase, it would have been part and parcel of whatever neural mechanism(s) mediated denial of mortality and reality, thus allowing the establishment and selection of a full ToM in human ancestors. Once this psychological evolutionary barrier was breached by the emergence of appropriate neural changes, there was no turning back. The generalized blunting of risk aversion that one sees in humans today would have been one of the side effects of this transition. At first glance, this would seem to be detrimental for survival, placing humans at greater risk from various dangers. However, full ToM—not only understanding of the minds of others but also having the ability to understand past, present, and future, planning ahead, and so on—came along in the package. Thus the blunting of risk aversion could now be put to positive use, allowing humans to take much greater (calculated) risks. And the benefits of doing so could be myriad, both for the individual and for the group.

How suddenly (in evolutionary time) did this transition occur? One of the ongoing debates in biology is whether evolutionary changes (including the emergence of new species) occur gradually (as suggested by Darwin) or in sudden fits and starts (as suggested by Stephen Jay Gould and Niles Eldredge in their concept of "punctuated equilibrium"[21]—a stable evolutionary state that suddenly changes). As is the case with most biological arguments, there is truth to both sides of the debate. A more accommodative view of these two ends of the spectrum is Sewall Wright's classic concept of an "adaptive landscape" of hills and valleys, in which a particular evolutionary feature has reached a fitness peak but cannot get across to the next optimal peak until it first goes through a valley.[22] The psychological evolutionary barrier that we have posited here may be thought of as a deep valley from which it is very difficult to climb out to the next peak. In other words, the cognitive evolution of warm-blooded birds and mammals may have proceeded through several peaks and valleys until it reached the optimal peak of self-awareness with rudimentary ToM. But getting across to the next optimal peak (full ToM) may have required going through a very deep and dark valley: the full recognition of reality and mortality.

Yes, we almost certainly reprioritized various unconscious emotions as we became more intelligent. However, the new and relatively sudden emergence of the imperative to value personal survival (or, more accurately, death avoidance) so highly could not be compensated for simply by a minor tweaking of our existing irrational behavioral drivers or our executive functions. Acquiring, maintaining, and benefiting from a full ToM required a novel strategy in order to overcome the negative consequences to natural selection. Will an intelligent being accept a 10 percent risk of the cessation of its existence in order to increase its reproductive fitness by 20 percent? Natural selection says to go for it, but this is not what any thinking individual with full ToM would do. Reproduction is fine, but not if the price is death. We suggest that many long-lived social animals have been repeatedly bumping up against this full ToM barrier for millions of years. So full ToM posed a new dilemma for evolution, and it is one that demanded a novel and unusual solution.

CHAPTER 6

Breaking through the Wall

Yea, though I walk through the valley of the shadow of
death, I will fear no evil: for thou art with me; thy rod
and thy staff, they comfort me.

—Psalm 23:4

Death, be not proud, though some have called thee
Mighty and dreadful, for thou art not so.

—John Donne, in *Holy Sonnets*

Some warm-blooded social creatures, such as dolphins,
elephants, chimpanzees, orangutans, and corvid birds,
likely have a form of "self-awareness," at least as defined by
studies such as the mirror self-recognition test. And formal
tests on some animals and birds have shown evidence of a
rudimentary ToM. For example, a study showed that a wild
chimpanzee who encounters a dangerous snake (actually a
plastic model placed there by scientists) will make a specific
alarm call only if other chimps in the group have not seen the
snake—indicating that the chimp had some knowledge that
the others have minds of their own.[1] Nicola Clayton and her

colleagues[2] have studied scrub jays that were allowed to hide away their food either in private or while being observed by another jay and then recover their caches in private. One group was then given the opportunity to steal the other birds' hidden food caches and the other was not. It turned out that thieves re-hid their own food caches if they were observed when first hiding the food. But the scrub jays that had no experience of stealing did not show the same subterfuge. Again, a reasonable interpretation is that the jays were both thinking ahead and reading the minds of the other birds.

Milind Watve and his colleagues[3] have also shown that the green bee-eater (*Merops orientalis*, a small tropical bird) can appreciate what a predator can or cannot see. Bee-eaters typically avoid entering their nests in the presence of any potential nest predator. Watve found that bee-eaters entered the nest more frequently when a simulated predator (a human experimenter) was unable to see the nest from his or her position, whereas they entered the nest less frequently when the predator was stationed at an approximately equidistant position from which the nest could be seen. And the bee-eaters entered the nest more frequently when the simulated predator was looking away from the nest. The human's angle of gaze from the nest was also associated positively with the probability of entering the nest, whereas the human's angle from the bird was not. A reasonable interpretation is that the bird has some understanding of the mental state of the simulated predator. Studies of crow and raven behaviors also suggest they have rudimentary ToM. For example, crows that are regularly fed by the same human will sometimes bring along colored objects to the benefactor, possibly as "gifts of appreciation."

But of all nonhuman animals, elephants may well be closest to the brink of attaining a full ToM.[4] It is a common saying that "elephants have long memories." These may not just be folk observations. Some unusual incidents have been reported in Africa following negative encounters of humans with elephants. The Amboseli National Park was set up to protect elephants in Kenya, but its establishment disrupted the way of life of local Maasai tribesmen. There have since been ongoing problems caused by elephants raiding and destroying the crops of the Maasai. Some Maasai have reacted by killing the elephants (young Maasai men were already used to demonstrating their virility by spearing elephants). Following these encounters, strange incidents occurred in which cattle owned by the Maasai were attacked and killed by the surviving elephants.[5] In the Maasai way of life, cattle are their most precious things, next to their own kin. We can of course never be sure what the elephant attacks mean. But since elephant attacks on cattle are unexpected, one possible explanation is that the elephants understood that the Maasai valued their cattle and that this was their way of indirectly getting back at their killers. In other words, instead of attacking the humans who killed their friends, the elephants seemed to exact revenge by killing the precious cows owned by the humans. If this is true, the elephants may well be on the brink of having a full ToM of the Maasai.

Other intriguing events occurred after the Pilanesberg National Park was created in South Africa. Some park organizers (assuming that young elephants had no knowledge of the personhood of others) decided that it would be easiest to just move the young elephants to the park but cull the older individuals, who were more difficult to resettle. This effort was carried out with the best of intentions and was based on available knowledge of elephants at the time. But the young elephants that observed the killing of their mothers and elders later grew up to become killers of other animals (particularly rhinos) in the new park. This bizarre behavior ended only when the park staff moved some adult bull elephants into the same area, restoring a sense of community to the younger elephants.[6] While even less straightforward than the case of the Maasai cattle, these events suggest that young elephants could remember the killing of their mothers and other elders and recognize this killing as having occurred to others like themselves. Regardless of the details, these and other observations[7] support the notion that this intelligent species has been bumping up against this *psychological evolutionary barrier* at various times.

The realization of one's mortality first requires fully comprehending the death of another individual. And this realization requires that one first understands the minds of others as individuals like oneself—in other words, it requires full ToM. If so, then other animals who do not appear to have a full ToM should not be fully aware of the death of others of their own kind. But in fact, the literature is rife with instances in which other animals do seem to react to death. Some reports indicate that chimpanzees witnessing the death of another of their kind may change their behavior in a manner suggesting that they did recognize the loss of a person.[8] Tetsuro Matsuzawa and others also point out that chimpanzee mothers whose infants die may carry the carcass around for quite a while,[9] as if grieving over the loss. Elephants in Africa will linger at the body of a dead pack member for a while, and are claimed to even return to the site where other elephants died to stroke and roll the bones of the dead one, as if to indicate that they recognize the death of that individual. After the death of a crow or raven, it is not uncommon for others to mill around for a while, cawing loudly. And of course there are stories of domestic dogs who show signs of grief when their owners (or companion pets) have passed away.[10] All these findings (only some of which are properly documented in peer-reviewed publications) suggest that many self-aware animals may have a glimmer of recognition of the death of others.

It is difficult to be sure if any of these are true examples of death awareness. Also, one does not know how commonly such behavior happens in a given species and whether or not we are hearing reports of exceptional cases only. But even if all these reports are accurate, it is quite clear that these forms of death recognition are nothing quite like what happens in humans. In our species, the death of a loved one is accompanied by very obvious and often vocal signs of grief that can last for a long time, along with attempts at comfort by others around the bereaved person. Depending on religious or cultural tradition, this is followed by various types of funerary rites, special memorial services, anniversaries, and so on. Thus humans are likely the only species that absolutely and fully recognizes the death of another individual to the point of being able to then transfer that reality toward understanding our own mortality. The existence of these anecdotes and case studies in which diverse other species seem to show various levels of grief *actually supports our theory*, as it indicates that some of them may be "on the brink" of understanding death.

This brings us back to the same stark question: Why haven't other species progressed to the point of fully understanding death the way humans do? It appears that many intelligent social animals have been on the brink of gaining a full ToM, but they've failed to reach the level of awareness that humans have. This actually favors the idea of the existence of a strong evolutionary barrier that has prevented these other animals from crossing over. And the best explanation is that they are unable to cross over because truly recognizing the death of another individual would result in recognizing one's own mortality, and the terror of this would be evolutionarily maladaptive.

All these animals could, in principle, gain advantages from being smarter. Yet only humans managed to break through this wall. This suggests that we have some special mechanism that compensates for the potential selective disadvantage we encounter when we place too much emphasis on survival purely for the sake of staying alive (as opposed to staying alive for the sake of reproduction). How do we deal with the contradictory demands of maximizing both our reproductive fitness and our odds of survival? The answer is quite simple: We just deny our mortality. Despite all rational evidence to the contrary, humans generally don't actually believe that they will die. At first glance this may seem a contradictory statement. Of course we all know we are going to die. We see evidence for this around us daily. The point is that despite our rational and intellectual understanding of death, we routinely deny the reality in a practical sense, on a daily basis, in the way we behave and function.

Granted, if you were to take a poll and ask people on the street if they anticipated that, someday, they would stop breathing and die, there is a good chance that one hundred out of one hundred would answer in the affirmative. However, a majority would also tell you that they didn't believe that this would mean the cessation of their existence. Many humans believe that following their mortal life there is some form of afterlife for their souls. We all agree that death is one of the two certainties of life (along with taxes), but we don't agree on what death implies. Living, breathing immortality is an impractical dream to most, but life ever after is not.

Certainly the most striking evidence of our denial of mortality can be found in religions. All human societies have religiosity of some kind, and most religious systems have some theory of an afterlife, be it reincarnation, going to heaven or hell, joining with ancestors, or something else. In other words, most humans gain reassurance concerning mortality from their religion. But while religion is likely part of the story, it cannot be the whole story. What about atheists? While the numbers of avowed atheists are still relatively small (between 10 and 20 percent of the US population, by some estimates), there are many people who don't call themselves atheists but still disbelieve anything supernatural and do not feel that there is anything more after they die. So if religion were a requirement for denial of mortality, all atheists should be constantly afraid concerning their mortality. But while a few may fear death,[11] the great majority do not. Of course, with some exceptions—such as the Cārvāka atheistic school of Indian philosophy, which traces its origins to 600 BCE, and the Sophists of Greece—atheism is likely a relatively modern phenomenon (some would consider the original form of Buddhism to be atheistic). Even societies with a sophisticated understanding of sciences firmly believed in their religion and their gods. Indeed, if someone in earlier times had sufficient information to be able to come to atheistic conclusions against a major prevailing religion, she or he probably would not have survived for very long if these views were made known to the high priests! Times have changed, and atheism is now an accepted way of life for some. However, it is interesting that the annual convention of the organization known as American Atheists is a rather small affair.[12]

All human cultures have some form of ritualized spiritual beliefs, which may be codified to varying degrees. Formalized religions often provide detailed descriptions of what type of experience one can expect postmortem, complete with rules for attaining entrance to various strata of society in the afterlife. This can take on markedly different forms. In some religions, one is reincarnated and returned to the earth as a new mortal entity. In others, specific destinations are assigned for life everlasting, depending on how well one behaved in mortal life. Some religions are more vague. What almost all religions share is a mechanism to reassure us about continued existence following the cessation of our physiological being—a denial of the finality of death. It would be interesting to further study the rare exceptions, such as Judaism, which leaves specific details of the afterlife somewhat vague,[13] to understand what other mechanisms compensate for not having a clear-cut belief about what happens after death.

When we think of religion, our minds typically go directly to the currently established major religions of the world, such as Christianity, Hinduism, Islam, Judaism, and Buddhism. But ancient civilizations also practiced various forms of piety in service of the same ultimate goal. The pharaohs of Egypt went to extraordinary lengths to surround themselves with accoutrements that would serve them on their voyage into the great beyond. Preagricultural tribes throughout the world today also have religious or spiritual beliefs that serve the same fundamental purpose. Indeed, if we want to ask when human ancestors first reached a level of intelligence that enabled full ToM, we should look for the earliest dates at which internments (burials) were accompanied by ornaments and appliances that suggest a future destination for the spirit of the corpse. We are limited by the physical preservation of such sites as well as by the interpretive conclusions of archaeologists—e.g., at some early burials it is unclear whether the internment was just a sanitary or social practice or whether it had religious significance. But human skeletal remains dating from around one hundred thousand years ago, found in the Skhul cave (located in what is now Israel),[14] were buried with various grave goods, including marine shells transported from many miles away and the mandible of a wild boar, which was found in the arms of one of the skeletons. The site also contains a joint grave of a woman and a child. By this point it is reasonable to suggest that those burying the dead were able to identify the buried individual as a living person who had gone through the transition of death.

With regard to burials by the Neandertals,[15] the situation is less clear. These extinct cousins of ours did sometimes bury their dead, but typically without funerary items. So it is not clear if it was simply a sanitary practice. And they seem to have begun to do this only about seventy or eighty thousand years ago, after more than one hundred thousand years of prior independent existence in Europe and western Asia. There is even a theory that Neandertals may have been copying the behavior of us humans (whom they probably first met around that time) without really knowing why they were doing it. We can of course never be sure about this claim. And there is new evidence saying that Neandertals may have been decorating themselves with bird feathers.[16] Overall, we just have to say that Neandertals were likely at the brink of gaining full ToM and may or may not have crossed this line to some extent (we earlier mentioned the fact that Neandertals sometimes took care of the injured and elderly). Steve Mithen, Thomas Wynn, and Frederick Coolidge[17] make a credible effort at using the archeological record to try to reconstruct the minds of these very close evolutionary cousins. Their general conclusion is that while Neandertals were likely similar in many ways, they were also rather different, especially in terms of their ability to relate to one another. The overall conclusion emerging is that a Neandertal would not have done well in a human society, or vice versa.

Hand in hand with our belief in an afterlife is our need to find a "meaning" for life itself. People want to think that they are "put on the earth" to serve some larger purpose; that they are somehow "part of a grander scheme." Why are we more comfortable thinking that we are pawns of some higher power(s) manipulating our lives for his/her/their enjoyment? Because the alternative—that there is no meaning, no scheme—also implies that there is nothing after death. The deep anxiety and depression generated by this latter conclusion has kept mankind searching for the meaning of life for thousands of years and will continue to do so for the foreseeable future. This is a common feature of human existential angst: Who am I? What am I doing here? What is the purpose of my being here? What is going to happen to me? Perhaps these age-old questions arise because even while we deny our mortality from moment to moment we also use our full ToM to think about and discuss these issues, at a depth that no other animal can.

There is no indication that the answer will be uncovered any time soon. Back when humans thought the earth was flat and infinite and that stars were holes in the dark sky, we didn't have any concept of DNA, cells, or the understanding of physiological life that science has provided. It was easy and even rational then to evoke spiritual entities in order to explain the inconceivable complexity of our being and to connect our consciousness to an indefinable soul that exists beyond our tangible flesh. Now that we have a much more realistic view of the world and our place in it, some humans no longer feel the need to accept that a higher being is running the show.

But even if one does not consciously believe in God or in gods, one could still be in profound denial of mortality. Many individuals have some sense of spirituality without accepting the formal rigors of an established religion. This essentially constitutes an individual religious construct and can satisfy the need for immortality of the soul as well as an established religion can. But whether one confesses to spirituality or not, there are some who will now assert rationally that they believe that death brings an absolute end to existence. But really, does anyone actually practice an absolutely atheist philosophy 100 percent of the time? Who among us has not, however rarely, made a little unspoken prayer in a time of stress? And is not irrational hope a form of prayer, addressed to oneself? Virtually all of us have at some time asked that a loved one be granted the ability to survive a medical crisis, win a contest, or some such thing. This is true even when the plaintiff has no firm concept as to the identity of the individual/spirit/deity who is being asked to look upon the situation with favor. You might say, "Yeah, I've done that, but it really didn't mean that I expected someone or something to be listening." Rationally, this may be true. But this behavior uncovers a deeper spiritual persuasion that lurks in the subconscious. As many other authors have written, this tendency to religiosity appears to be a consequence of our evolution, and there may even be a brain region mediating some of these kinds of religious thoughts.[18] Even if you have found an alternative to religion, you have simply entered an alternative form of denial. But if religion were required to assuage our fear of mortality, then atheists who abandoned religion should all have an overwhelming fear of their mortality. This is not the case for most atheists, suggesting that there are more hardwired mechanisms at work in such denial. On the other hand, confirmed atheists tend to be highly educated individuals with much knowledge, perhaps allowing them to rationalize their mortality. Of course, as one gets older there can be a greater appreciation of the notion of death, as well as acceptance that it awaits. And given the low reproductive potential of middle-aged and elderly people, natural selection would no longer be an operative factor. But even the old person who has accepted the inevitability of death does not brood on it all the time. Reality denial is still there, as a comfort!

If one were to fully and continuously contemplate one's existence and the repercussions of its end, it would lead to constant anxiety, stress, depression, and paralyzing behavior in many ordinary circumstances. Obviously this is not a problem for most individuals, fitting the notion that we have a primary mechanism for denial of reality and mortality that was evolved for this purpose. But if this denial arose because of a discrete and localized neural module that was specifically evolved only in humans, then there should be instances in which this module has been damaged by localized strokes or brain tumors; individuals with this condition would then develop a constant and *exclusive* fear of death. Such a person might be said to have "isolated thanatophobia" (Thanatos is the Greek demon that personifies death). Yet such a syndrome does not seem to exist as an isolated condition (individuals who have a constant fear of death typically have many other phobias and anxieties, and thanatophobia is just one of them).[19]

So, as suggested earlier, the mechanisms involved are more likely to include a partial degradation of a preexisting subconscious mechanism for the recognition of death risk.

Most people seem to go about their lives as if mortality is not an issue. We are in denial of the implications of our ultimate mortality. Whether overtly religious or not, virtually everyone is willing to risk death, even if the risk is small, in ways that do not make sense in a perfectly rational world. We smoke cigarettes, despite overwhelming evidence that this behavior will likely lead to a premature death. We refuse to wear seat belts, knowing full well that they will increase our chances of survival in an accident; never mind the fact that just getting in the car is dangerous to a degree. We risk a fatal disease by practicing unprotected sex. We eat the wrong foods, eat too much food, and don't get enough exercise. We agree, and even volunteer, to become soldiers and fight for our country. It's not just religious or otherwise spiritual people who engage in these activities. Irrational risks are taken by practically everyone, including individuals who would assert that, consciously, they do not believe in an afterlife. Typically, we rationalize this behavior by saying that we are weighing the risks against the benefits, the latter often being that the behavior makes us happy. But is a little happiness worth an increased chance of nonexistence, in a totally rational assessment? In reality, we are expecting that it will be someone else who becomes a statistic—another symptom of our capacity for denial. Regardless, these behaviors demonstrate that we are also in denial of the implications of our ultimate mortality—indeed, we are in denial of denial itself. And this is all part of our broader penchant for denying reality, for which there is even more evidence all around us.

This reality denial at a deep level of our being is essentially universal, and is a fundamental quality of being human. Perhaps even more telling than its pervasiveness is the degree to which this denial exists in many people. A fundamentalist Christian must believe that some of the reality that we know about the world from science is not true. The earth is less than ten thousand years old. Dinosaurs did not exist (or, if they did, they lived side by side with humans). The physical laws that we accept as inviolate in every other facet of our lives do not apply when God is involved. Of course, no one has actually seen these laws violated recently, but centuries ago someone wrote that they were, and so it must be true. In short, a fundamentalist Christian is required at one time or another to deny things that his or her experience in the real world says is true. The fact that so many adherents of so many established religions accept so many physically impossible scenarios speaks volumes about the massive capacity of humanity to exist in a constant state of denial.

There is an intellectual disconnect here that is almost impossible to bridge. Literal readings of the creation stories of any religion are difficult to reconcile with the large body of incontrovertible scientific facts about the origin of the universe, the solar system, the earth, and life on earth. So many people in the modern world have a difficult time with absolute acceptance of the doctrines of fundamentalist religions. But they find accommodations that allow them to accept much of modern learning while maintaining their basic faith and their belief that nature is God's handiwork. This still requires some degree of reality denial; it's just a question of scale. A compromise position is taken by the famous physician-scientist and evangelical Christian Francis Collins in his book called *The Language of God.*[20] Collins suggests that while all well-documented scientific facts are true, the probability of the Big Bang having sustained itself, the probability of life having appeared on earth, and the probability of the emergence of moral humans via evolution from such life are all so highly unlikely that these can be seen as indirect evidence for an almighty God. I once had the opportunity to challenge Collins in person as to how he reconciled his faith in rigorous scientific exploration with his belief in God. His answer was that if one truly believed in an almighty God one should not have to worry that God was staying awake at night concerned about the feeble attempts of humans to understand his creation! The reader interested in this attempt to bridge the divide is encouraged to read both the Collins book and Francisco Ayala's *Darwin's Gift to Science and Religion,*[21] which addresses the fundamental dilemma of theodicy: Why does a just, omnipotent, and merciful God allow so much suffering to occur? If you accept the compromise position that God set evolution in motion and did not interfere thereafter in the process, then you have your answer to theodicy: Evolution (set in motion by God) requires suffering in order to succeed. And, as Ayala reminds us, the Catholic Church accepted this position a while ago.

The attacks on evolution that currently preoccupy many fundamentalists provide an interesting insight. Over the centuries, the advance of knowledge based on scientific inquiry in the Western world has created conflicts with various teachings of Christianity. Galileo's persecution by the Catholic Church over the position of the earth in the cosmos is the most famous of these. Centuries later the Church relented, and the Vatican now runs an astronomical observatory. However, people find it more difficult to make compromises between faith-based teachings and science when it comes to evolution. Why is evolution "a line in the sand" from which there can be no compromise? Is it because evolution strikes at the very heart of the reason people seek religion? One can make an allowance for the solar system as an entity that is described only metaphorically in the Bible. It doesn't jeopardize the idea of an afterlife; you just have to travel a bit farther now to get to heaven. However, if one accepts the scientific notion of evolution by the processes of random mutation and natural selection, man's uniqueness is threatened, as are any notions of a higher meaning to life, the immortality of the soul, etc. Of course, as Francis Collins points out, there is an intermediate position that allows for full acceptance of evolution by natural selection without necessarily eliminating the possibility of a higher meaning to life.

The scientific community does not deal very effectively in communicating with the public on some of these issues. Thus, even though the scientific evidence for evolution is every bit as overwhelming as the evidence for the organization of the planets, the latter can be embraced by everyone—yet some people can never accept evolution. Similarly, people often feel threatened by those who practice religious beliefs different from their own. Indeed, history reveals countless examples of religion being used as a justification for the slaughter of nonbelieving infidels—situations in which "nonbelievers" are typically defined as those who hold different beliefs from our own, rather than no beliefs at all. Why is this? Many formal religions require absolute acceptance of their tenets for successful transition to the next life. Perhaps the existence of another possible form of the after-death safety net threatens the reliability of the one on which one has come to rely. If someone else's religion has validity, then we are at grave risk ourselves. Destroying the practitioners is a way of destroying the threat. Is this why we care so much whether someone follows the same spiritual path as we do? We don't know, but we do tend to be more tolerant of non-religious social and cultural differences if they don't directly impinge on our own beliefs and lives.

After this seemingly harsh discussion on religion, we feel the need to insert a conciliatory note here. The text above may come across as an antireligion polemic—that is definitely not the intent. It is just impossible to reconcile science with the views of fundamentalists who insist on absolute faith in myths that cannot be verified. On the other hand, there is plenty of room for compromise and even common ground among individuals who accept all the scientific facts yet leave room for spiritual approaches to things that cannot be explained by science. Books such as those of Ayala and Collins attempt to address these issues, as do others, including Kenneth Miller's *Finding Darwin's God*,[22] Michael Ruse's *Science and Spirituality*,[23] Louis Perry's *Thank Evolution for God*,[24] and some of the previously mentioned books about human religiosity.

Holding a full ToM requires a large dose of mortality denial in order to compensate for the selective disadvantages that accompany the "personal survival first" mentality. Religion supplies a formal device that can satisfy this requirement. Unfortunately, some of our more popular religions have evolved into large, structured, political and economic entities that go well beyond the purpose for which they were initially created. This often leads to religion in general getting a bad rap. But many religions and other spiritual constructions have managed to provide a comfortable degree of denial without being overly intrusive on everyday life. Nor does the acceptance of the central necessity of denial as the driving force for the development of religion prove that there is no God, or that all spiritual beliefs are a load of bunk. Science can no more disprove the existence of a higher being than it can prove that such an entity is watching over us. Evolution does not mean there is no plan; it just puts constraints on what the plan might be. Accepting the existing knowledge base concerning evolution certainly does not in any way rule out the possibility of an underlying plan that initiated evolution. As we said earlier, while we have a pretty good idea of how life on earth evolved over hundreds of millions of years, no one can say with confidence how life originated. Scientists have performed experiments to show how organic life may have been created, but if you want to invoke the hand of God in this event, no one can prove you wrong. It seems reasonable to simply accept that spirituality is a denial-essential quality of humanity and to try to make our spiritual sides as congruent with reality as our individual psyches will permit. Indeed, Robert McCauley points out that while accepting religion comes naturally to humans, accepting the unexpected facts revealed by science does not.[25] And Paul Bloom suggests that religion is a by-product of the fact that children have a natural tendency toward a dualistic theory of mind.[26] Jesse Bering has even conducted interesting experiments with children to document such tendencies.[27]

It is particularly fashionable to castigate the Vatican (the headquarters of Catholic Christianity) for its historical refusal to accept new scientific discoveries, the persecution of Galileo being the best-known example. However, less well appreciated is the fact that the Vatican has a scientific council (the

Pontifical Academy of Sciences) that systematically evaluates scientific evidence and collects information to be used in future decision making (we already mentioned that the Vatican accepted the fact of evolution a while ago, seeing it as the best explanation for the problem of theodicy). Of course the official position of the Vatican is not always taken soon after a scientific council report. But they do take the information into account in future policy making. Do you know when it was that a multidisciplinary group of scientists first agreed that, six million years ago, humans and chimpanzees had a common ancestor? It was at a Vatican Pontifical Academy meeting in 1982. This meeting was described by J. M. Lowenstein in a *Nature* article cleverly entitled "Twelve Wise Men at the Vatican."[28] Our colleague Russell Doolittle was one of the twelve, so we have a firsthand description of the event from him (Doolittle also recounts that, in that sexist era, women scientists were not officially included—but in fact the note taker for the meeting was the well-known scientist Adrienne Zihlman of the University of California at Santa Cruz). While the common ancestry of humans and apes was undisputed among scientists, it was easy for religious doubters to criticize the overall conclusion—because there was much uncertainty and even acrimonious debate about the timing of the proposed common ancestry. The difficulty was that paleoanthropologists were relying on fossils of an ancient ape they thought was related to the common ancestor, and they had therefore dated the divergence time to more than twenty million years ago. Meanwhile, studies of protein sequences that used the so-called molecular clock were giving numbers closer to between five and seven million years ago. It was at this 1982 Vatican council (convened specifically to address this issue) that a group of experts from all sides first came together and concluded that the molecular data were more accurate—a conclusion that the general scientific community later accepted.

A more current example is that of climate change. The Pontifical Academy meetings on this topic have involved experts from all over the world, and eventually resulted in the pope making recommendations that jibe with scientific information on the subject.[29] It is particularly ironic, then, that the fiercest antiscience attitudes and denial of scientific realities such as evolution and climate change come from some of the Protestant Christian groups in the United States.[30] After all, the word *Protestant* came from the notion that this newer branch of Christianity was protesting the highly orthodox and excessively literal biblical orientation of the Catholic Church and its resulting practices. Perhaps it is time to "protest" against those who now seem to be in total denial of reality. It is quite one thing to deny evolution and thus damage the teaching of biology and the understanding of disease and other physical processes. But it is quite another to deny the reality of climate change. Here we are talking about a very dangerous planetary experiment in which humanity *has only one chance to get it right*. The major concern is that we are not doing so, and that it will be too late to turn back. More later on this critical issue.

The recognition that we humans are full of denial can be traced back to many philosophers and scientists, most recently Freud and his disciples. However, past explanations for this phenomenon and its consequences have been clouded in theories and ideas that have not only been difficult to prove but also have not stood the test of time. Potentially related to denial is the psychological state called depression, which is very common in humans and thought to be one of the most poorly diagnosed and treated diseases in the world. When someone has a temporary "situational depression" it is usually straightforward: We can understand why depressive thoughts arise from a recent negative incident or difficult personal situation. However, this does not explain the very common problem of unprovoked major depression,[31] which can even end in suicide. While depression is undoubtedly one of a variety of disease processes that are driven by various biochemical changes in the brain, major depressive disorder (as it is officially called) should be highly maladaptive in terms of survival and reproductive fitness. Among the many theories for explaining this disease is one called depressive realism.[32] As Tali Sharot puts it, "While healthy people are biased toward a positive future, depressed individuals perceive possible misfortunes a bit too clearly. While severely depressed patients are pessimistic, mildly depressed people are actually pretty good at predicting what may happen to them in the near future." Thus major depressives may in fact be the true realists who fully appreciate the enormity of all the negative issues that face them every day, socially, personally, and professionally. In this line of reasoning, "normal" people are the ones who are deluding themselves!

A surprising treatment advance has been the finding that ketamine (a short-acting anesthetic that is commonly used for animal and human surgeries) can have an immediate positive impact in some patients with depression.[33] If our hypothesis about depressive realism being a lack of reality denial is true, then further studies of the mechanism of ketamine action may provide clues as to how the optimism bias persists in nondepressed humans. Notably, ketamine is also a commonly abused drug, altering reality in a manner that many apparently find pleasurable. Ketamine can produce short-term hallucinatory effects at subanesthetic doses, followed by a sense of detachment from one's physical body and the external world, effectively enhancing one's state of reality denial. So perhaps ketamine is acting to counter depressive realism by enhancing reality denial.

Meanwhile, it is interesting that patients with prior episodes of major depression can sometimes enter a bipolar "manic" phase, in which they become wildly optimistic, suffer from a severe lack of understanding of reality, and do things that no normal person would consider reasonable. Such manic behavior might well be considered an extreme form of reality denial. Perhaps it is a mental backlash to the depressive realism that the same person had been through earlier.

Panic disorder is another common psychiatric condition, characterized by anxiety attacks in which otherwise mentally healthy individuals have sudden feelings of panic and danger and even think that they are going to die.[34] Those who experience such episodes may end up in the hospital emergency room, thinking they had a heart attack or some other major medical problem, only to find a bit later (when the attack passes) that everything is fine. The underlying neurobiological mechanisms of panic disorder are largely unknown, but evidence suggests that the amygdala, a brain region that plays a key role in the neural network of fear and anxiety, is central to the process (we will hear more about the amygdala later). In the context of the theory espoused in this book, one can suggest that these anxiety attacks represent a sudden episodic failure of the human reality denial system, transiently unmasking the fear of death. The fact that there is no known *naturally occurring* equivalent of this disease in other animals (there are artificially induced models) suggests that it is telling us something about human cognitive origins.

Returning to depression, it is notable that the most serious complication is suicide—a behavior that appears to be uniquely human.[35] The myth about lemmings jumping off a cliff in mass suicide turns out to be based on unsubstantiated folklore, which was propagated by an early Disney movie in which lemmings are actually thrown off a cliff by humans. On the other hand, there are some reports of dolphins in captivity doing things that might be considered similar to suicide.[36] Another instance may be that of whales, who sometimes beach themselves for unexplained reasons and die despite valiant efforts by humans to save them. And there are cases in which aged animals are said to wander off from their group, apparently going away to die. It's hard to know what this means, but it seems likely that the individual who leaves is exhausted, unable to keep up with the group, and simply takes the path of least resistance. Indeed, if other animals were truly capable of suicide, this would most likely happen among those kept in long-term, caged captivity, under very depressing circumstances. But while such animals might pine away by refusing to eat, clear-cut examples of deliberate animal suicide in captivity are hard to find. In contrast, incarcerated humans often attempt suicide, and prison officials take great care to ensure that the opportunity does not present itself.

Overall, there is very little solid evidence that any of the large-brained, apparently self-aware animals and birds we have discussed in previous chapters ever take their own lives intentionally. Human suicide, on the other hand, is remarkably common, and is prevalent in all societies (although extremely rare in children under the age of seven or eight, prior to full ToM and understanding of mortality). In order to commit suicide you have to know what death is and that you are capable of inducing your own death. In some cases suicide can be attributed to very difficult circumstances that an individual is faced with—personal, professional, or financial. However, a significant number of suicides occur in individuals without such obvious explanations, but are instead suffering from major depressive disorder.

Overall, one can suggest that suicide may be an outcome of having a full ToM without adequate reality denial. Note that from the evolutionary perspective suicide is generally maladaptive. If you kill yourself, you will fail to contribute your genes to the next generation through new progeny and you will eliminate the possibility that you can help any existing younger progeny to survive. Of course suicide can also take the form of a deliberate act in which a human is trying to make a point or achieve a goal. Examples of this range from the drugged Japanese kamikaze pilots of the Second World War who went to their certain deaths by flying planes into enemy ships, to the current spate of suicide bombings associated with certain terrorist groups. In the latter instance these individuals are neither depressed nor have difficult circumstances forcing them to commit suicide. Rather, they are deluded into completely denying reality and truly believe that their actions are for the greater good of their own cause. And in some cases they truly believe that there is a reward waiting for them after their suicidal martyrdom.

After discussing a depressing topic, let's interject a "lighthearted" story about suicide. I once saw a patient in a coma in the emergency room. The patient had attempted suicide many times before and failed. More than once (as in this instance), he had taken an overdose of the very antidepressant pills prescribed for his depression. But it is difficult to commit suicide with such pills. In the 1970s, when this incident occurred, an overdose of the pill in question (amitriptyline) would basically put you in a coma but was unlikely to kill you unless you developed an abnormal heart rhythm. Thus the standard protocol for such patients was to pump out their stomachs and wait for the drug's effects to wear off. But because of the risk of a heart rhythm disturbance, such patients would also be on a cardiac monitor. At 3:00 a.m. I was called back to the intensive care unit because the patient's rhythm was unstable. In that era, the correct response was to administer physostigmine. This antidote works very fast and can even wake the subject in a short time. After the injection, I watched the patient stirring, following which he sat bolt upright. He looked at me with a crestfallen face and said, "Oh, no, not again!" You probably found this story amusing despite the grim underlying theme. But this is just an example of our ability to make fun of death in many different ways (more on this later).

Another way humans can effectively commit suicide is to deliberately place themselves in circumstances in which the only certain outcome is death at the hands of other humans. Perhaps the most famous example is the story of Jesus of Nazareth, who, fully recognizing the consequences of what he was doing, not only put himself in great danger but may also have taken active steps to ensure that he was apprehended. According to the biblical story, Jesus knew the outcome and was going through it to fulfill his godly mission. Whether or not you believe in the specifics, the story is so compelling that it forms the core of one of the world's largest religions. This in turn led to the martyrdom of many early Christians, who delivered themselves to certain death by publicly proclaiming their beliefs. The most well known of these martyrs was the disciple Peter, whose upside-down crucifixion laid the groundwork for what is now called the Roman Catholic Church. Many other early Christians went to their deaths at the hands of the Romans and other non-believers, doing so willingly in a form of intentional suicide because they believed that their reward would be paid in another dimension. Remarkably, less than three hundred years later, the entire Roman Empire adopted Christianity as its primary religion.

In keeping with the above discussion, several lines of research and related books converge on the notion that reality denial is a dominant feature of human thinking and behavior. We already mentioned the studies by Tali Sharot and her colleagues, which show that humans are optimistic even in the face of negative information. Sharot has shown that neural pathways originating in the frontal lobe of the brain may be related to this form of irrational thinking.[37] This work was influenced by the studies of Dan Gilbert, who has shown that we humans generally tend to have a basic level of happiness that we revert to eventually. We also tend to find ways of rationalizing unhappy outcomes so as to make them more acceptable to ourselves. In fact, events and outcomes that we initially dread often turn into new opportunities for happiness when they do actually take place.[38] Dan Ariely (*Predictably Irrational*)[39] and others have also written about the fact that humans routinely make decisions that seem to go quite against rational thinking, even when they've been provided with all the necessary information. Another related line of research has to do with the evolution of overconfidence (believing you are better than you actually are). A priori, one might imagine that overconfidence is a bad thing, as it is likely to get you in trouble. However, Dominic Johnson and James Fowler have shown by theoretical evolutionary modeling that reacting in an overconfident manner can actually have fitness benefits, as long as the contested resources are sufficiently large compared to the cost of competition.[40] And, counterintuitively, seemingly sensible unbiased approaches are only stable under limited circumstances. Regardless, what is overconfidence but the denial of reality? It seems that once reality denial has settled in, overconfidence can actually be a successful approach. Human history shows us many examples in which this turned out to be the case. Of course, there are many cases in which overconfidence resulted in massive failures, such as the economic meltdowns of 2008, when businesspeople made trades and bets that were optimistic beyond all reasonable expectation. Much has been written on this sorry episode in the world economy, for which we are all still paying the price.[41] Unfortunately, in the years that followed, the reality denial has continued, both within governments and in the business community!

A corollary to our overall hypothesis is that religiosity and religious explanations for what happens after death may have partly evolved as an adaptation to full ToM and awareness of mortality. If so, individuals with autism spectrum disorders (ASDs)—whom we described earlier as having a limited ToM—might manifest a lesser tendency for religiosity. Available studies support this notion,[42] and it is said that deeply religious parents of children with ASDs sometimes complain that they are unable to get their children to appreciate their religious beliefs. Of course, one difficulty stems from the social communication deficit these individuals suffer from. But, as Jesse Bering points out, in the autobiographies of high-functioning autistic individuals, "God, the corner-stone of most people's religious experience, is presented more as a sort of principle than as a psychological entity. For autistics, God seems to be a faceless force in the universe that is directly responsible for the organization of cosmic structure—arranging matter in an orderly fashion, or 'treating' entropy—or He's been reduced to cold, rational scientific logic altogether."[43] A related observation is that individuals with autism are less prone to hypocrisy, apparently because (unlike "typical" people) they are less concerned about what others think of their behavior and actions.[44]

Having considered many wide-ranging issues and facts that impinge on our overall theory, let's move on to discuss how the transition past the proposed psychological evolutionary barrier might have actually occurred. Isolated denial of reality should be a *risky* thing for any animal to have. And we have also argued that isolated theory of mind should be psychologically *unstable*. If both theory of mind and reality denial are each individually bad for fitness, how, then, could evolution have possibly gotten *two immediate negatives* to work together at the right time, in the right order, allowing a *permanently positive* outcome?

CHAPTER 7

How Did Reality Denial Emerge?

All war is based on deception.

—Sun Tzu, in *The Art of War*

When one is in love, one always begins by deceiving one's self, and one always ends by deceiving others. That is what the world calls a romance.

—Oscar Wilde, in *The Picture of Dorian Gray*

Up to this point, we have made arguments mostly based on fairly firm experimental or logical footings. From here on, some aspects of our discussion will be deliberately speculative. For example, from where did our enormous capacity for denial of mortality and reality arise? Imagine the time when one of our ancestors first became intelligent enough to achieve a full ToM and then grappled with understanding the reality of mortality. The first human in this condition could not just say: "Golly, this death thing really sucks. I had better figure out a way to avoid acknowledging its implications or at least thinking about it anymore. Otherwise, I will make decisions that will inhibit my ability to reproduce—not to mention the fact that I'll be racked with anxiety." There are many reasons it wouldn't have happened this way, in addition to the obvious one—that such sophisticated logic had not yet evolved. First of all, this level of logical analysis would have been a bit much at this stage of the game—we weren't nearly as smart then as we are now. And there is a catch-22 here: How does one begin to logically deny awareness of mortality without first having evolved the ability to analyze it? This denial does not necessarily have to reach the massive proportions that our current intelligence level requires, but it had to be sufficient to deal with an awakening sense of reality and mortality.

Most important, evolution does not usually evolve a trait after it is required. Fully terrestrial reptiles did not evolve from fish that just decided to flop up onto dry land and have a go at breathing air. A primarily aquatic creature first evolved a mechanism to gain oxygen from air in order to supplement its gills, allowing it to take advantage of spells outside of the water. This tactic became more developed over time, permitting longer periods of exploration on dry land. Eventually, lungs resulted, getting the animal past the physiological barrier. How did the original evolutionary processes that generated humans get past the analogous psychological barrier? A full ToM would have presented an acute challenge—like suddenly placing a fish out of water—requiring a dramatic adaptation in order to counteract the negative effects of the survival-versus-fitness conundrum. A species cannot develop a full ToM and then later evolve denial of mortality in order to solve the resulting problem. There must be at least some rudimentary degree of denial already in place in the population. Following this logic, the initial emergence of reality denial must have preceded emergence of full ToM. However, this does not necessarily imply a new brain mechanism. It could well be that preexisting mammalian brain pathways for automatic subconscious fear of dangerous situations were simply tuned down or even damaged and perhaps culturally buffered. Such tuning down of fear responses might have allowed emergence of full ToM and awareness of mortality.

Regardless of the mechanism, *reality denial likely had to be waiting for us when we first achieved a full ToM.*[1] But it would be risky for any animal to deny or ignore reality, so this phase could not have lasted very long. When thinking about the frequencies at which events occur to support evolution, quite rare events can often suffice. Mutations in any given gene happen at rates that may seem glacially slow but are more than sufficient to provide the genetic diversity necessary for evolutionary events that take thousands of years. However, even when considering probability on evolutionary time scales, what are the chances that a species will have developed a sufficiently strong capacity for reality denial (a detrimental feature, initially) independently of its need for deflecting the anxiety attacks arising from mortality awareness? And since there is no evidence for a single gene mediating such changes, this would have had to happen in more than one individual at the same time, so that they could then mate with the other. Altogether, this was an extremely unlikely event, and probably only occurred once in the history of the planet. The improbability of such denial emerging *just prior* to the emergence of full ToM may be the reason why it took so long for even one species to become truly intelligent.

If the human brain had to be programmed for denial before attaining full ToM, what was the evolutionary driving force for this? There is no compelling evidence to answer this question. It's like the question of how life first arose: For such one-time events we can mention possible answers, but we can never say with conviction what actually happened. Denial of mortality would not leave any specific trace in the fossil record. However, the related awareness of the deaths of others could be manifested by the emergence of death rituals, which first occurred around one hundred thousand years ago. In this regard, we have already mentioned examples of the other early behaviors we attribute to modern humans (such as making bead necklaces) that also require full ToM and multilevel intentionality. Archaeological evidence for such "modern" human behavior coincides reasonably well with genetic estimates of the timing of the emergence of our species. It might be that our proposed mechanisms came into play only at this time, coinciding with (and being critical for) the final emergence of our own species. More on this matter later.

Let's return to our speculative scenario. Our ancestors are getting smart enough to be tickling the full ToM barrier. They are living in relatively stable social groups with established social hierarchies. Individuals could communicate in fairly sophisticated ways, and something at least approaching language (protolanguage?)[2] was developing. Does this language need to be more sophisticated than the communication methods of present-day primates? We don't know, but probably. And, as is true for current humans, there was likely significant investment in the raising and training of children from both parents. But biologically, eggs are far more costly than sperm in most animals. That is, human reproductive rates in the population are very sensitive to the number of eggs that mature and are fertilized to become embryos, babies, and the next generation of adults. Sperm, on the other hand, are very cheap, and only a very lucky few will ever find an egg and have a future. Assuming that relatively monogamous human pair bonding already existed in humans at the time,[3] the supplier of sperm is likely going to be a contributor to food and shelter and potentially to the raising of offspring. This makes mate selection especially critical for human females. They are biologically programmed not only to seek a mate who is healthy and has strength, speed, and other survival skills but also to seek one who will be a good father as well.

In this intraspecies competitive environment (society), each male will be trying to convince females that he will be a good choice as a mate. He will seek to demonstrate good physical traits, excellent abilities as a provider, and good intentions with respect to future behavior. Females will be evaluating the candidates with a careful and suspicious eye. The male is desperate to be selected as a mate: It is an event that increases his fitness more directly than almost any other. In this situation, males are known to puff up their résumés. Up to this point, such processes of sexual selection and mate choice do not actually have to be conscious. But if it can influence the female choice favorably, selection will also favor males who are good liars. Wonderfully explicated in the writings of Robert Trivers,[4] the germ of this particular idea relates back to Charles Darwin's implicit hypothesis that sexual selection and mate choice were involved in the emergence of human cognition. Darwin's classic book *The Descent of Man*[5] is effectively a companion to another one called *Selection in Relation to Sex*, which lays out his theories about sexual selection. In this confusing tome, Darwin appeared to be saying that sexual selection was much involved in human evolution, but he didn't detail exactly how. Now we can suggest that the ability to hold a false belief and even believe one's own falsehoods could have had a positive selective effect on mate choice in humans. And if this trait also allowed persistence of full ToM and increasing utilization of intelligence, then it would have further favored mate choice among such individuals.

The traditional idea that human intelligence evolved primarily because people chose mates for their general mental abilities is difficult to reconcile with classical sexual selection, in which there is usually a big disparity between the two sexes—such as the peacock's tail, which only the male displays. In the case of humans, there is no evidence for a major difference between male and female mental abilities and capacities for rational thinking (though many long-suffering women might argue with this statement!). In humans, sexual selection and mate choice based on improved cognition is possible in both directions. It is plausible that the ability to be a better liar would have increased mating possibilities in an environment where some degree of language and complex communication were already present. This of course begs the question of how such language and complex communicative behaviors could have evolved without the prior emergence of full ToM and multilevel intentionality. As with other major transitions in evolution, such as the appearance of the first aquatic creatures who had developed the ability to breathe air and live on land, many critical things would have had to happen all at once to get past this difficult barrier.

Lying males could potentially get the best mates. Of course, the opposite sex does not want to succumb to lies— they want the genuine article. This means that there will also be selection for good lie detectors, especially among the females. So when it comes to mate choice, there could have been a two-way cognitive arms race. Males would be selected to be better and better liars, and females would counter by being selected to be better and better lie detectors. Evolving something approaching a full ToM would be very helpful to both sides in this mental gamesmanship. The effects of these traits on fitness are especially obvious in mate choice, especially considering the intrinsic biological inequality of eggs and sperm. Any lies that assist in the acquisition of resources will also be selected for. Such resources can be tangible entities, such as food, or intangibles, such as status in the social hierarchy, that lead indirectly to the accumulation of food, shelter, and so on.

So even in a relatively simple society with rudimentary language skills, being a successful liar will improve one's ability to influence potential mates and others, and one will be better able to compete with one's neighbors. It will increase fitness and will be selected for. Conversely, knowing when one is being lied to is a good thing, and it will also be selected for. Both of these traits are well developed in any good poker player; at the top level they're what the game is all about. But they are also used in life in general. There is a sophisticated literature on cheaters in complex social systems— literature suggesting that the ability to detect and punish cheaters could be positively selected for[6]—but this is something we will not go into here.

In the lying arms race there is one ultimate weapon, a tactic that can circumvent even the best lie detector. This strategy can fool a champion poker player and the coyest socialite. No matter how subtle your "tells" might be (to use the poker parlance), no one can detect your bluff. How can you lie without ever tipping your hand? You simply believe the lie yourself. If you completely convince yourself that whatever nonsense you are spouting is actually true, no lie detector can pick up the behavioral or even physiological signs of untruth. If what you are saying is actually part of your personal reality, there are no telltale signs that can give you away. This ability to deceive oneself seems to be a uniquely human trait and possibly a requirement for deceiving others.

Thus when competing for resources (especially mates) within even a primitive social framework, there is a selective advantage in making oneself out to be better than one is in reality. It is also advantageous to be able to recognize when a member of the group is being less than honest. However, if one believes one's own lies to be the truth, it is impossible for another person to detect even subtle behavioral signs of lying. This will provide a selective advantage for those who develop a controlled capacity for self-deception. Of course, we aren't trying to make a case that becoming a habitual liar is a good thing. As is the case with all traits, there is a cost as well as an advantage. For the cost of self-deception to be tolerable, it must be moderated and channeled in ways that can provide the maximum benefit at a minimum price.

In making this argument, we have been referring to the development of self-deception as a mechanism that improved our ability to lie and therefore to influence others. But this self-deception is just one form of reality denial, a form that can develop into more comprehensive denial skills with relative ease. That is, our need to deny the realities of death could have had its beginnings in traits that were selected to improve our fitness by making us better liars. We recognize that there is a bit of circular argument here. In order to be a really effective liar one needs to have full ToM and multilevel intentionality to begin with. It is also possible that the two abilities (denial of reality and self-deception) simply fed back on each other relatively quickly over a short period of time to achieve the current human condition. And if they positively affected mate choice and reproduction, this could only help to rapidly "fix the trait" genetically in succeeding generations. Of course it is likely that many genes were involved, which also were affected by our interactions with the environment.

We can only speculate as to how complex our communication skills were, how accomplished we were at deceiving others, and how much self-deception (and what kind) was useful at that critical time. It is not hard to imagine that as we developed the first inklings of the true consequences of death there was just enough denial in place to put these anxieties aside. All we had to do was deny death risk a little bit—not enough for our intelligence to dissuade us from doing things that maximized fitness. As we got a little smarter, and a little more cognizant of death, we were able to ramp up our denial abilities accordingly to divert our anxieties. At that point, denial was serving a new selective purpose. Thus denial of mortality is likely just part of a general human ability to deny obvious realities even in the face of evidence, and this would change the way denial continued to evolve. As we became smarter and smarter, we also got better and better at denying reality, specifically when it suited our purposes. Optimism and overconfidence could now kick in, with their benefits and their attendant risks.

The difficult part was just getting through the early stages, when there had to be enough self-deception to support a level of denial that allowed us to do things not in the best interests of our individual survival (as opposed to our survival as a population). And it was extraordinarily improbable. Many intelligent animals probably bumped up against this barrier for tens of millions of years before all the traits came together in sufficiently elaborate fashion at the same time to allow one species, us, to deflect the fitness costs of full ToM. But once that happened, we could increase our denial ability along with our smarts, eventually reaching the extraordinary heights on both fronts that we currently experience. Whether or not becoming more accomplished liars was the driving force for the original development of denial independent of full ToM is unknown. What is certain is that we now employ denial to deal with the unthinkable consequences of our reality and our mortality.

Of course, the very first time the potent combination of reality denial and theory of mind emerged in a small group of our human ancestors, it might not have been psychologically very stable, as two additional factors could have wreaked havoc with this already delicate situation. As we discussed, having a full ToM allows one to be much more effective in lying to others. And given the positive value of lying in the race to obtain mates, there might have been *ongoing positive selection for the ability to lie*. Thus it could well be that at this early stage lying was extremely common, especially between potential sexual partners. Unlike the story line of the popular movie *The Invention of Lying*, which imagines a time when lying was first invented, the initial problem may have been *too much lying*. This would have further destabilized the already unstable combination of reality denial and theory of mind, creating a chaotic situation for the social group. The only solution might have been mechanisms to actively suppress lying. While parental and societal punishments work today to achieve this goal, such approaches may not have been feasible at the outset. Rather, there may have been active selection for mechanisms to suppress lying. This possibly explains why most humans feel guilty about lying and even have detectable physiological responses when doing so—except for the rare psychopaths discussed by Simon Baron-Cohen who can escape detection of such responses by a lie-detector test.[7] As Steven Pinker suggested: "Mind-reading may in fact comprise two abilities, one for reading thoughts (which is impaired in autism), the other for reading emotions (which is impaired in psychopathy)."[8] A second negative factor operating at that time might have been a high rate of suicide among those who had come to understand their mortality but had not yet attained the full denial of this reality. As Nicholas Humphrey suggests: "Any tendency to suicide could not but pose a serious threat to the continuity of the human species. In fact, if unchecked, it would certainly have been terminal." So the rapid up-ramping of the optimism bias may have been critical at this stage. Remember also that this is all happening in a small population of human ancestors who have just developed this unusual combination of abilities and have barely begun to be positively selected for the genetically wired stability of the combination of reality denial plus theory of mind. While *this is all pure speculation*, such a complex series of events is plausible in the evolution of the human mind. Regardless of the validity of the speculations, this massive capacity for denial is now a fundamental quality of being human. Perhaps Charles Darwin's famous double volume *The Descent of Man* and *Selection in Relation to Sex* could be retitled *The Descent of Human Denial, with Selection in Relation to Sex* (a suggestion made by Pascal Gagneux).

So far we have cited many facts and ideas that appear to support our theory and have not found any that negate it. But since we cannot conclusively prove the overall thesis, we must continue with our "one long argument" in favor of it by citing many more clues and additional examples of reality denial. In later chapters we will address why our ability to deny reality has enormous and potentially disastrous consequences for humanity *as well as much potential positive value.*

CHAPTER 8

Evidence for Reality Denial Is All Around Us!

A man is never more truthful than when he acknowledges himself a liar.

—Mark Twain, quoted in *Mark Twain and I* by Opie Read

We always deceive ourselves twice about the people we love—first to their advantage, then to their disadvantage.

—Albert Camus, in *A Happy Death*

The human penchant for denial of reality is actually so obvious that it is surprising how long it has taken us to fully acknowledge it. Stop and think for a moment about your place in the cosmos. There is now strong evidence that this vast universe began more than thirteen billion years ago in an event known as the Big Bang.[1] There is also evidence that time and space are vastly more complex than we can perceive,[2] and that we may well be in one of numerous parallel or alternate universes.[3] Our Milky Way galaxy is just one of at least one hundred billion galaxies and our sun is a minor star among more than one hundred sextillion stars in the universe (that's a 1 followed by twenty-three zeros—or one tril-lion times one hundred billion). Our own Planet Earth is just a minor piece of rock within our own puny little solar sys-tem. Life on earth (our biosphere) exists within a thin layer of gas (the atmosphere), within the oceans and other bod-ies of water (the hydrosphere), and within a very superficial crust of the earth (the lithosphere). All this life originated perhaps only once, remained rather simple for more than two billion years, and eventually evolved into complex life forms, including some with rather sophisticated mental abilities. But all of them were still unable to truly understand the reality of their situation—until humans came along.

As a modern-day educated human capable of understand-ing all the above, you now know that you are nothing more than a sort of synthetic "hologram" composed of subatomic particles that originated from the Big Bang, which, when brought together in a particular combination, generated a brain, a mind, and a conscious state—which in turn give you the introspection that allows you to read this book and un-derstand it for the very, very, very brief space of time that you are alive. Why is it that despite understanding how abso-lutely, totally, and completely trivial you are in the enormity of this cosmic reality you can step out every day and act as if the universe is your oyster? Why are you not terrified and de-pressed by the fact that you are just the tiniest little transient blip in this enormous reality of time and space? Those of a religious bent will simply answer that faith in God is what al-lows it. But this ability to deny reality exists even in those who are atheistic and appreciate that they are nothing more than the previously mentioned hologram.

Consider also your own daily experience. You will see that we are indeed very good at convincing ourselves (and others) of things that actually may not be true. In particular, we tend to be influenced by how many times we are told something. Advertisers realize this. Repetition has a remarkable ability to get us to accept certain ideas, such as which car constitutes the best value, which cola tastes the best, and so on. As I write this, we have recently "survived" the December 21, 2012, Mayan day of apocalypse, which many otherwise sane people actually believed would be the end of the world as we know it. Some of these people even made elaborate preparations for the event. Of course this is just the latest in a series of such apocalyptic predictions that have been made during the history of humanity. In a December 2012 survey conducted by the Public Religion Research Institute in partnership with the Religion News Service, more than one-third of Americans said they believed that the severity of recent natural disasters is actually evidence that we are in what the Bible calls the end of days.

But it seems that the entities who take fullest advantage of the power of repetition are governments. For example, most Americans truly believed that the 2003 invasion of Iraq was a legitimate response to the terrorist attacks of 9/11 and that the nonexistent "weapons of mass destruction" were real.[4] History is also full of instances in which denial of realities by national leaders proved to be disastrous. A classic example is Neville Chamberlain, who was UK prime minister from May of 1937 to May of 1940. Against all the obvious evidence, he continued to sincerely believe that Adolf Hitler's aggressive activities could be contained by an "appeasement foreign policy"—which included the 1938 Munich Agreement, in which he conceded portions of Europe to Germany. It was only when Hitler continued his aggression by invading Poland that Great Britain finally declared war on Germany, on

September 3, 1939. By that time Hitler had built up his war machine to the point where it almost overwhelmed Britain. Nazi propaganda chief Joseph Goebbels later said, "If you repeat a lie often enough, it becomes the truth." Some modern "news" channels and politicians in the United States have taken this approach to the level of an art form, especially when it comes to election-related rhetoric. As Bill Clinton put it on *The Daily Show with Jon Stewart*, US politics is now characterized by "fact-free fighting."

We are particularly susceptible to this type of self-deception when the lie is so attractive that we really want to believe it. Politicians are adept at promising simultaneous tax cuts and increased government services. Of course, once enacted, the tax cuts are frequently followed by deficits and reduced services. But the promises still seem to be effective to varying degrees come Election Day. Meanwhile, self-deception works at various levels of our own consciousness. As anyone who has engaged in heated political arguments can attest, we are always sure it is the other person who is obviously deluded in his or her analyses of facts and events. As Robert Kurzban puts it, everyone *else* is a hypocrite.[5] But even someone who rationally knows the score can also be deceived at a deeper level, subconsciously. And this can have a profound effect on behavior. We may acknowledge all the overwhelming statistics, but it's the "other guy," not we, who will die from cigarette smoking or drinking too much.

Note that it is not just repetition from another source that effectively turns fantasy into reality. We convince ourselves of rationally absurd notions by self-repetition as well; this is the essence of the development of self-deception as we are proposing it. And it is not just politicians who become accomplished liars by deceiving themselves. The numerous bizarre conspiracy theories on issues ranging from the JFK assassination to governmental cover-ups of alien visitations to the perpetrators of the 9/11 attacks are all the more remarkable because large numbers of intelligent people actually believe them. Such a list could go on and on, as Michael Shermer has so eloquently demonstrated.[6]

Even scientific researchers can become wedded to theories based on relatively limited evidence. Although the primary goal of the scientist is to seek the truth about reality, there is a tendency to believe in one's own theory—sometimes even against contradictory evidence. This is why scientists are asked to make hypotheses that are testable, so that they can check whether or not they are true. Of course, completely avoiding speculation is not a good idea, either, as important concepts may only be partially testable today yet later become more verifiable. However, one often finds that the more an individual espouses an idea, the harder it is to convince him or her otherwise, even in the face of contradictory experimental evidence. This type of false validation by repetition is a far more common problem in science than the very rare experimental fraud. It is also harder to detect, as the proponents of the false theory are true believers. Intelligent, educated people are no more or less susceptible to such repetition than anyone else; it's an integral part of being human. Ironically, you can level some of these very same accusations against this book you are now reading. We have chosen to "hang our hat" on a theory that is not fully testable and to focus on all the information that supports our theory. But we have also looked hard for any facts that would negate our theory, and have so far found nothing.

The most obvious evidence of reality denial can be found in the sociopolitical and religious domains, and it would be easy to present many more examples. However, such realities change with time and the examples might well be dated by the time you read this book. So we will confine our remarks to the two polarized positions in 2012 American politics. Psychologists such as Jonathan Haidt[7] have written eloquently about the underlying mental processes that may make individuals more or less "conservative" or "liberal" in the US political system (these terms have somewhat different meanings in a historical context, and also in some other countries). In brief, Haidt suggests that there are six moral foundations that all humans relate to, and that liberals and conservatives place disproportionate emphasis on certain of these foundations (he demonstrates that it is liberals who are not paying enough attention to some of these key foundations). Haidt, along with a group of social and cultural psychologist colleagues, has presented an extended moral foundations theory—"to understand why morality varies so much across cultures yet still shows so many similarities and recurrent themes." The theory proposes that several innate and universally available psychological systems are the foundations of "intuitive ethics."[8] Space does not allow a full discussion of this matter, but the fact that some of these intuitive but incorrect beliefs persist in the face of scientific evidence supports our notion that denial of reality is a fundamental human characteristic.

Regardless of the history and the details, it is clear that neither "liberal" nor "conservative" positions in US politics truly recognize reality. In other words, both views are in denial of reality. (Ironically, if you try a Web search on the term *denial of reality*, many top hits are either US conservatives or liberals, each accusing the other of denying reality!) In fact, the best solutions are likely to be found by acknowledging reality and taking the appropriate middle position on the major issues. Of course, such middle positions are the hardest to hold, as they can be attacked by both extremes.

A classic example is the ongoing debate about marijuana (cannabis) in the United States.[9] The cannabis plant has been with humans since prehistoric times, at least in some cultures. Besides its well-known mind-altering properties, cannabis has long been used for medicinal purposes of various kinds (although its effectiveness is mostly scientifically unproven). In the early part of the twentieth century, cannabis was even a legal part of the pharmacopoeia in the United States, and doctors could prescribe it. However, following the 1933 lifting of the prohibition against alcohol in the United States, there was a large federal government organization (the Federal Bureau of Narcotics) that no longer had enough work, and they went looking for something new to focus attention on. Largely for this reason, cannabis (renamed marijuana) was not only criminalized but also demonized as a "killer weed," and an active attempt was made to characterize it as a much more serious problem than it had ever been. Fast-forward to the 1960s, and "pot" became the signature drug of the counterculture. As with alcohol during the Prohibition era, attempts to eliminate cannabis actually resulted in an increase in its use, and eventually made it into a favored revenue source for South American drug cartels as well as illicit growers in the United States.

At the moment, political views about marijuana lie at two extremes. At one extreme we have a view that inappropriately puts it in the same category as much more dangerous and highly addictive drugs, such as heroin and cocaine. This categorization results in a massive and expensive interdiction program as well as unnecessary incarceration of many otherwise reasonable and nondangerous people—not to mention drug-related violence across the United States–Mexico border. Recently, South American drug cartels have decided it is not worth the trouble of smuggling this bulky drug into the United States. So the single largest site of marijuana production is now thought to be in the Sierra Nevada mountain range of California. Illegal immigrants are dispatched by South American drug lords (sometimes under threat of harm to their families) to set up temporary camps for marijuana cultivation on the mountain slopes. These hapless individuals have everything to gain and nothing to lose by carrying out their mission, and they are often armed with guns. This in turn requires major ramping up in the numbers of drug enforcement personnel, who have to find, confront, and stamp out these growing sites. A side effect has been the risks for hikers who are unfortunate enough to stumble onto one of these camps. Meanwhile, there is the ongoing effort of the federal government to eradicate local production, even when state law approves marijuana for medical reasons. All this serves to drive the price up further and elevate the risks associated with the whole activity.

At the other extreme are those who feel that cannabis is a totally harmless drug, incapable of causing any addiction, and that it should be freely available to anyone who can grow it in their backyard. This view is also outdated. It is now emerging that chronic long-term marijuana smoking can have negative health consequences, and can trigger the onset of diseases such as schizophrenia in those at risk.[10] And while adolescents who think cannabis is harmless are initiating its use at ever-younger ages and are even smoking it daily, there is increasing evidence that it can have a permanent neurotoxic effect on their brains.[11] Meanwhile, given the high frequency of marijuana's current illicit use, there are also likely many "stoned" individuals driving on highways, operating heavy machinery, and doing other things that really require one to be of a clear mind. Of equal concern is the success of breeding programs for improved strains of marijuana (partly a response to criminalization and pressure from law enforcement). There are now cannabis strains tens of times as potent as the old "weed" that used to be smoked in the 1960s. Thus any opinions about cannabis safety originating from someone who only "smoked pot" in the 1960s are out of date.

These two extreme positions represent the majority of people who take a stand on the matter (and others listen in bemusement when they hear reports that marijuana is now the number one cash crop in the United States). Neither extreme is accepting reality. Both groups are denying the realities we have just mentioned in regard to the dangers of their respective positions. The realistic and practical solution would be to legalize cannabis but control its production and sale in a manner similar to the way we control alcohol. Besides requiring licensing of production, we could levy a heavy tax on sales. We could also require that individuals who use cannabis be held to the same kind of standards we have for alcohol today. As with alcohol, age limits for legal consumption could be set based on the latest research about detrimental effects on the developing brain. If you're an adult, what you do in the privacy of your own home might be up to you, but you cannot become a danger to others in a public space. And if you hold a position of responsibility— operating heavy machinery or doing other jobs in which your mental alertness is vital—then it would be illegal to be high on cannabis at your job, just as it is illegal to be drunk.

Added benefits to this realistic approach would be a huge increase in tax revenue for cash-strapped states, better control of quality, and greater transparency about potency for the consumer. Warning labels could be put on cannabis products, similar to the labels on bottles of wine and distilled spirits, that would allow individuals to titrate their consumption to the circumstances. Doctors could then freely ask their patients about cannabis use and counsel those who should not be using it. Medical research could also better uncover additional effects of cannabis use on health, positive or negative. And while true addiction may not occur with cannabis use, a form of psychological dependence on the drug does exist, and treatment programs could be instituted for dependent individuals.[12] Last but not least, this approach would pull the financial rug out from under many South American drug cartels, even while saving taxpayers a lot of money on interdiction programs that would no longer be needed.

This realistic and sensible approach avoids most of the risks of the two extreme positions. Although it has its own limitations, the approach could also be modified in the future based on the latest research, and any problems that arise would be easier to manage because they would be out in the open (as they are now with alcohol). But, as with all such issues, it is much easier for most people to deny reality and comfortably stake out one or other of the extreme, unrealistic positions. As of this writing, though, experiments somewhat along the lines suggested above have been approved by voters in the states of Colorado and Washington.

Let's move on to the additional evidence of denial that can be found in the very common risk-taking behavior of humans, whose curiosity—like that of some domestic cats—can cause them to take unnecessary risks. However, humans seem unique in the extreme extent to which we indulge in risk-taking behavior. Examples are easy to come up with, such as engaging in warfare under circumstances in which death seems very likely and exploring extremely dangerous territories where the chance of survival is low. Other examples of extreme risk taking can be found in the initiation rites of some tribal societies or modern fraternal organizations, especially for males. Of course, some of these kinds of behaviors might be explained by human social pressures and/or by the possibility of gaining something in return. It is harder to explain examples of voluntary serious risk taking that have no value other than the pure thrill of the risk itself. Modern examples include bare-handed scaling of sheer cliff walls, white-water rafting on extremely dangerous rivers, extreme bungee jumping, skydiving, surfing in very rough waves, and the dangerous British practice of jumping off coastline cliffs into uncharted waters where the depth and risk is unknown ("tombstoning"). While many of these behaviors involve a significant death risk, those who practice them routinely rationalize the risk. And there are many examples of individuals who have nearly died from such activities yet then return to do the same thing again, assuming that such an accident will simply never happen twice.

Here is an example from personal experience, presented with permission from a friend. This young woman finally met the man of her dreams while she was participating in recreational skydiving. He was a very experienced jumper and had already suffered through one major skydiving accident in which he broke his back, consequently suffering long-term injuries and chronic pain. But he had of course gone back to his thrilling and risky sport. They finally decided to get married, buy a house, and have children (he even agreed to reverse his vasectomy for her, so she could have his children). Just as they were preparing to complete landscaping of the house and adopt a cat, he decided to participate in a "swooping" competition, an extreme form of skydiving sometimes called canopy piloting—something he did quite frequently. This is a low-altitude form of parachuting in which the contestant executes high-performance maneuvers and comes in for a landing at very high speed while swooping dangerously close to the ground, demonstrating skills in speed, accuracy, and distance. That particular day, my friend's husband's primary chute had a malfunction at high speed, and, sadly, he died when he hit the ground. Despite his vast experience, why did he not open his emergency chute, which would have saved his life? While it is possible that he was already unconscious when the primary chute malfunctioned, the autopsy showed no evidence of a heart attack or stroke while he was in descent. We will never know the truth, but it is possible that he was denying the reality of his extremely dangerous situation. Regardless of exactly what happened, here was an individual who had previously survived a serious accident resulting from the same kind of activity. Hc had just gotten married and was starting out a new life. Despite this, he went back to an extremely risky activity, which eventually cost him his life. And his widow recently talked proudly of an acquaintance of hers, Dan Brodsky-Chenfeld, who, despite having survived life-threatening injuries in 1992, went on to become a six-time world champion skydiver and has made more than fifteen thousand career jumps to date. Many more such examples can be cited from various risky human activities, but the point is already clear.

Of course, such supercourageous (or foolhardy?) individuals are relatively rare. What about the more ordinary humans who still subject themselves to similar risks? They do this even in the face of dire warnings, such as this typical waiver that a novice parachute jumper signs at Perris Valley Skydiving in California after seeing a video about the risks:

> By signing this document you state that you understand that by making a tandem parachute jump you risk serious injury or even death caused by any reason including design or manufacturing defect, malfunction or misuse of the parachute system, or improper careless, negligent, or even grossly negligent use of that system by the parties involved. Simply put, signing this waiver demonstrates your willingness to accept the risks of serious injury or death in exchange for the thrill of making a tandem parachute jump.

At some other "drop zones" (as these skydiving sites are called), each frame of the warning video includes an image of the Grim Reaper, to make the risk very, very clear. But when asked about the rate of dropouts resulting from such warnings, a manager at one of the sites replied, "It's rare . . . I can think of maybe two in the last few years. So a very low percentage—less than one thousandth of 1 percent."

Less extreme but equally puzzling is the risk-taking behavior we all indulge in every day (some of which we have mentioned earlier). This evidence for our facile reality denial is all around us. Let's begin at the personal level. Those living in educated societies are well aware of what needs to be done to take care of our health: avoid smoking, limit alcohol consumption, maintain an optimal body weight, exercise frequently, avoid unnecessary stress, eat more vegetables and fruits, avoid excess red meat, and so on.[13] But we don't bother with most of these obviously beneficial behaviors. We also forget (or even refuse) to wear seat belts and motorcycle helmets, or we indulge in drunk driving and texting while driving, activities that we know can lead to fatal accidents. And anyone who knows of the high traffic fatality rates on crowded roads in India and China is denying reality when he guns a flimsy motor scooter or rides a bicycle in between all those zigzagging cars.

We all know the old adage "An ounce of prevention is worth a pound of cure." But in the past we knew relatively little about optimal ways to prevent disease—the advice we got was mostly in the form of "old wives' tales" passed along by our grandmothers. Well, it turns out that a lot of what Grandma advised was actually right on the mark. We now have ample scientific evidence that there are many ways in which we can markedly reduce the burden of diseases. Although the recommended prevention might depend on the part of the world one is talking about, let's just look at so-called Westernized diseases. The big killers of the day in developed countries are cancers and cardiovascular disease (heart attacks and strokes). It turns out that the ways to prevent these diseases are quite well known,[14] and we have just mentioned some of them above. With regard to some risky behaviors, such as smoking, there is at least the personal excuse that addiction makes it difficult to break the habit. Not so in these other instances, in which appropriate behavior is all that is called for. Yet the vast majority of us fail to follow these simple preventive measures. We know what to do as individuals and as a society, but we do little to follow through. Instead we face the increasing burden of diseases of a preventable nature and the skyrocketing costs of paying for the consequences, individually and as a society. Perhaps you are less aware that breast-feeding has a very positive effect on the health and eventual IQ of an infant and that it reduces the incidence of breast cancer in mothers.[15] But modern societies make it very difficult for a woman who wishes to have children and breast-feed if she has a job outside the home. Indeed, combined with the fact that breast and ovarian cancer risk increases substantially when a woman does not have children and does not breast-feed,[16] one can argue that our current working environment is carcinogenic for women. Thus I once wrote an opinion piece regarding the "leaky pipeline" in which women of childbearing age "mysteriously" drop out of potentially successful careers in academia: "It's [lack of] Proximate Childcare, Stupid!"[17]

While progress has been made with smoking reduction, this was achieved mostly by raising taxes on cigarettes, by promoting disgust for smoking among nonsmokers, and by emphasizing the risks of secondhand smoke. We are at least past the era when American cardiologists and lung specialists were the ones in the hospital who seemed to smoke the most cigarettes (though some still do not exercise regularly!). However, smoking rates are increasing in developing countries, and in China, nearly 30 percent of all male cardiologists still smoke.[18] On a broader public health level, we are well aware of what we need to do. There is convincing evidence that spending money on disease-prevention programs rather than on expensive cures gives the taxpayer a bigger bang for his or her buck. But despite all the evidence, those who try to encourage public health and disease prevention have the hardest time getting cooperation from individuals, from the general public, and even from the government. It has taken decades of public education to make even some limited progress in such matters.

Much closer to home, I can cite an example of educated scientists who deny the reality of the very thing they are working on in their research. In my own lab, I study the mechanism by which the eating of red meat (beef, pork, and lamb) results in the well-known increased risk of heart attacks, cancer, and early death.[19] We have discovered a molecular mechanism that may explain the association, and we hope that this knowledge can one day be used to mitigate the risk. I recently conducted an anonymous poll among members of my lab group to ask whether their full knowledge and close working relationship with the molecular reality of a major health risk had altered their personal behavior. The poll results showed that around 70 percent of the group had been moderate to heavy red meat eaters prior to joining the group. All these individuals responded that they understood both the epidemiological and molecular information and fully agreed with the nature of the risk. However, despite this, only one-third of them had significantly reduced their red meat intake, and only one person had quit altogether. Here again is the remarkable ability of humans to deny reality in the face of the facts, even when they are intimately involved in studying those very facts. During a recent trip I was even sent a photograph of the entire lab group thoroughly relishing large quantities of grilled ribs! So they were not only denying the reality of their increased death risk but were also mocking the possibility.

At a family level, we are also in denial. We know, for example, that we should do all things necessary to ensure proper inheritance planning, have adequate life insurance, and so on. But most of us put this off, even considering it a distasteful subject for discussion. We know what we should be doing at the family and social level to maximize our mutual cooperation, but we rarely try. Denial of reality is also obvious at the national level, with regard to both politics and the economy. Most countries (until the recent crises) were actually hoping that their economic problems would simply go away. For that matter, how many American citizens know what the level of their national debt is and what the real meaning of that number is? If you really want to know, check one of the websites that track the numbers in frightening real time.[20] As of this writing, the total US national debt is more than sixteen trillion dollars (that's the number 16 followed by twelve zeros!). Each man, woman, and child in the United States thus owes more than fifty thousand dollars to US Treasury bondholders throughout the world. And since only about half our citizens pay any taxes, the actual average owed per taxpayer is more than one hundred thousand dollars (and even more for people in the highest tax brackets). It is interesting to note that when the national debt reached one trillion dollars (around 1980) there was much wailing and gnashing of teeth in the media and in society at large. The angst was less when we crossed the three-trillion mark, and the problem essentially disappeared from public view after that, so we blithely went past the ten-trillion mark without any fanfare. And despite political platitudes about "not ruining the future for our children and grandchildren," nothing meaningful is really being done about this problem. Instead, Americans take comfort in the fact that many other developed countries are even worse off in terms of the ratio of their national debt to their gross national product!

Indeed, we even tend to deny the most fundamental aspects of our origins and history as a species. In *The Story of B*, a 1996 novel by Daniel Quinn,[21] the protagonist, B, talks about the Great Forgetting of Civilizations. He points out that most of us have forgotten that there was a time when people were sustained primarily by hunting and gathering rather than by animal husbandry and agriculture. By the time we invented writing and history began to be recorded, it was simply assumed that people had come into existence as civilizations. B argues that our knowledge and worldview would be greatly altered if the great thinkers of our culture had known back then that there was history before the beginning of civilization. While archaeology has now uncovered the facts, our worldview still remains unaffected. But Daniel Lord Smail has taken up the challenge,[22] arguing that our true history began when our brains evolved to their present-day capabilities—in effect, from the time of the evolutionary transition described in this book.

We are also very good at conveniently forgetting about the risk of natural disasters. Many years ago, I completed my medical training at Washington University in Saint Louis and was fortunate to be offered a faculty position there. For various reasons I wanted to move to the West Coast and decided instead to accept an offer at UC San Diego. Among the lighthearted gibes this generated from my colleagues in Saint Louis, the commonest was "Why would you want to move to earthquake country?" I decided to look into the matter and was relieved to find that San Diego is actually an area of California in which serious earthquakes have been quite rare. But to my great surprise I found out that Saint Louis was close to the epicenter of one of the worst earthquakes in recorded US history! Between December of 1811 and February of 1812, the New Madrid Fault Line[23] unleashed a series of earthquakes that shook the entire region for almost three months. Although the Richter scale was not in existence at the time, estimates suggest that at least three tremors would have been measured at a magnitude of 8.0. It was reported that church bells rang hundreds of miles away, as far as Boston and the Carolinas. The earth split in many places, forming large fissures and "sand volcanoes." Large-scale earth movements resulted in the Mississippi River running backward, generating massive flooding and a new lake. These cataclysmic events are well recorded and are the subject of ongoing investigations. In fact, current data suggest that such earthquakes may occur every few hundred years, and this region of the Midwest may be in for another big one. Worse still, many major gas supply lines run through this area, generating the likelihood of a major fireball following another quake. Estimates of the consequences of a repeat event run to hundreds of thousands of lives lost and tens of billions of dollars' worth of damage. Surprisingly, most of my colleagues in Saint Louis had not even heard of the New Madrid Fault Line or of these massive earthquakes. If they had, they consigned them to the category of a historical anomaly that would never happen again—not a matter to worry about. This exemplifies how we humans are very good at forgetting catastrophic events and just blithely returning to our regular way of life. Meanwhile, geologists have been warning community and political leaders in the Midwest about the risks of the New Madrid Fault Line. Despite this, construction codes and disaster preparations in Saint Louis and Memphis (the city at greatest risk) remain quite inadequate to handle events of the kind that happened between 1811 and 1812. Of course, when geophysicist Seth Stein of Northwestern University began to suggest that the risks were exaggerated,[24] he got much more attention.

Other examples abound. The great influenza pandemic of 1918 killed more people than both world wars combined—

and yet until recent times (when the subject was brought up in popular books[25] and by the scientific community) this catastrophic episode was practically forgotten. And of course there are many examples in which humans return to rebuild in the shadows of active volcanoes or in places where earthquakes, flooding, and deadly hurricanes are frequent. Our denial mechanism simply allows us to suppress these horrific memories. We can wrap up all these examples of denial in the common Hindi phrase *chaltha hai*—which literally means "so it goes" and is also used as shorthand for a certain careless approach to life and flexible attitude to rules.

As we already mentioned, humans even enjoy "laughing in the face of death." Even though we recognize our mortality, we make jokes about it all the time. For example, the San Diego County Fair has a popular Heart Attack Café, complete with garish advertisements for "Coronary Burgers" and "Stroke Sundaes"—making fun of the fact that this kind of food could not only shorten your life but also result in disastrous consequences for your family if you indeed succumbed to a heart attack or a stroke. But the subject is felt to be quite amusing, and few would think twice about the seriousness of what is being implied. Another example can be found at the University of California, Berkeley. It turns out that their big football stadium was constructed right on top of the Hayward Fault Line, which is highly prone to earthquakes. In fact, the two halves of the stadium sit on either side of the fault line, which runs lengthwise under the middle of the football field. Since 1923 the walls at the north and south ends of the field have slipped past each other by about thirteen inches (thirty-three centimeters), and obvious cracks in the wall graphically illustrate the location of the fault. The consequences of earthquakes in this area are well known and can be disastrous, resulting in major loss of life. Despite this, the locals (including the academics) proudly point out to visitors that the stadium lies on the fault line, and everyone finds it quite amusing. Such human foibles are easy to find, from serious, life-threatening circumstances all the way down to common vernacular phrases that make fun of the possibility of death, such as, "You're killing me!" Even a human who is actually dying can make black humor out of the situation. The writer Oscar Wilde is said to have commented on his deathbed: "My wallpaper and I are fighting a duel to the death. One or the other of us has to go." Sir Thomas More apparently asked his executioner to help him up to the gallows, saying he'd be able to come down by himself just fine. And when Voltaire was asked on his deathbed to forswear Satan, he apparently quipped: "This is no time to make new enemies." So it appears that we humans not only deny our mortality but also joke about it freely, without engendering any sense of fear. Of course, some psychologists might argue that such jokes are an attempt to cover up the fact that we are subconsciously afraid of death. Whichever way you put it, it is reality denial.

In 2011 the world lost one of its great intellectual giants to an early death caused by cancer. Steve Jobs was the co-founder, creative genius, and visionary behind Apple computer company. Although the precise details are unknown, published information indicates that Jobs may have hastened his own death by denying reality.[26] It seems that when he was first diagnosed with a rare form of pancreatic cancer, the tumor might have been curable by surgery, because it appeared localized to the pancreas. Given his situation in life, it is likely that Jobs had access to the best expert advice and information and knew he had a chance for a permanent cure by surgery. However, he instead apparently chose to pursue various forms of alternative medicine, none of which had been proven to cure this particular kind of cancer.[27] By the time he finally went back to the surgical option, it was too late, as the tumor had spread beyond his pancreas. No further treatments could save his life. Why did someone as intelligent, well read, and knowledgeable as Jobs not follow what established medical science told him would be the rational and realistic way to *maximize his chances for a cure*? Was he in denial of the situation he faced? While we will never know for sure, this may be yet another example of reality denial when faced with the cold, hard facts. One simply has to look through history books to find numerous other examples of highly intelligent and rational individuals ignoring reality and taking unnecessary risks.

Let's now go even closer to home—to Danny and I. Earlier, we mentioned Ernest Becker, who wrote the important book *The Denial of Death*, which won the Pulitzer Prize in 1974. Becker actually died of cancer just before this award was to be presented to him. And, as you know, one of the authors of this book also died while trying to complete his story about denial of mortality. It is a good thing that the author still living is not superstitious! Actually, the fact that I can make a joke about my own death (and you may be chuckling about it yourself) exemplifies how we humans are able to make fun of mortality. On a more serious note, I hope I get this book published before I die—because if I do not, human superstition and irrationality could well inhibit all future work on this subject.

The Magic of Reality by Richard Dawkins[28] eloquently emphasizes that reality (i.e., that which is truly real and being continually revealed by science) is itself more magical than all the false realities that our minds construct. Read this book if you really want to appreciate what reality is. On the other hand, not all of us can fully understand and appreciate this reality while remaining undisturbed by it. After all, as we said earlier, you are nothing more than a kind of hologram made up of subatomic particles that come together to form atoms, which come together to form molecules, which in turn form the organs of your body, including your brain, which in turn generates your mind, which is what you're using right now to understand this reality—for only a fleeting period of time, in cosmic terms. And the daily realities facing many of us are far from pleasant. It is highly unlikely that any other animal can truly appreciate reality at this level. But denying reality helps us tolerate the ugliness of that very same reality.

Having speculated on how human reality denial might have evolved and then presented the evidence for such denial all around us, we can now move on to further consider the personal, societal, national, and global consequences of this remarkable, useful, and yet highly dangerous ability.

CHAPTER 9

Too Smart for Our Own Good

A syllogism: *Other men die, but I
am not another; therefore I'll not die.*
                    —Vladimir Nabokov, in *Pale Fire*

Those who cannot remember the past are condemned to
repeat it.
—George Santayana, in *The Life of Reason I: Reason in
                    Common Sense*

As we learn more about other primates, cetaceans, elephants, and corvid birds, we can see that they are
capable of far more incisive and complex thinking than was
once believed. Still, no other animal's mind can approach the
abilities of the human mind. This is obvious in the machines
we build, the art we create, and the conventions we construct
to organize ourselves into groups for work, play, and war. We
generate abstract ideas and complex communication systems
that can convey enormously intricate thoughts and exquisite
subtleties. We can think about the past and future, talk about
people who are dead and gone, about individuals yet to be
born, and even imagine the minds of fictional characters. In-

deed, it appears that our brains are outfitted to go well beyond the needs of basic survival in a competitive world. Looking back at our relatively humble origins as a species, we can see that the human brain appears to have long ago evolved remarkable capabilities that go far beyond what we needed for survival and reproduction at the time during which it evolved. Indeed, we are still continuing to find new ways to use our remarkable minds as technology advances and new ideas emerge. But how could natural selection have achieved this feat?

While we all know of Charles Darwin and his many contributions to science, we hear much less about the co-discoverer of the principle of evolution via descent by natural selection. As we mentioned earlier, Alfred Russel Wallace independently came up with the idea after a long and successful career as a naturalist, apparently while in the midst of a malarial fever fit in Indonesia. Darwin did the honorable thing and ensured that their seminal work was published side by side. But Wallace later lost favor among scientists, in part because he stated that he could explain everything about biology by natural selection *except* for the human mind. This, together with some other unorthodox positions that Wallace took on scientific and nonscientific issues, caused his legacy to be diminished.[1] Sadly, Wallace has even become an unlikely hero for the intelligent design movement, which has misunderstood his point about humans and distorted it to suit their thinly veiled creationist agenda.

Wallace was a flawed genius, getting some things quite wrong but others quite right. He discovered what is now called Wallace's line, which marks the deep oceanic separation of tigers and monkeys in Indonesia from kangaroos and koalas in Australia. This finding not only assisted him with his original thinking about natural selection but also helped originate the field of biogeography—the study of the distribution of species, organisms, and ecosystems in space and through geological time. But Wallace also pointed out that it is difficult to explain how conventional natural selection alone could have "preselected"[2] the human brain for the myriad novel functions it carries out today, a problem I have referred to as "Wallace's Conundrum." While the conundrum has been misunderstood and ignored for a long time, the challenge remains unanswered. Wallace asked how it is that one could take very young children from "primitive" tribes, move them to a situation of much greater opportunity in England, and elicit behavior from them that one would expect from locally born Englishmen. With our present-day knowledge of human origins, we must agree with and even extend Wallace's point, recognizing that the original humans who left Africa (as well as those who stayed behind) already had the innate ability to learn everything that humans can learn today, from cuisine to calculus, from computing to cricket. Wallace challenged his own basic theory, saying that he did not understand how natural selection could have "preselected" for such human abilities well ahead of the time when they might be useful. He asked, "How could 'natural selection,' or survival of the fittest in the struggle for existence, at all favor the development of mental powers so entirely removed from the material necessities of savage men?" And he concluded that the human mind was "an instrument...developed in advance of the needs of its predecessor."[3] In the same vein, Nobel laureate and early molecular geneticist Max Delbruck's book *Mind from Matter?* discussed the emergence of human brains and wondered why so "much more was delivered than was ordered." The great Indian philosopher

Swami Vivekananda addressed similar issues in his analysis of evolution, saying "Darwin's theory seems true to a certain extent. But...the struggle theory is not equally applicable to both kingdoms. Man's struggle is in the mental sphere." Like Wallace, Vivekananda says that the human mind is somehow different and not simply evolved by conventional natural selection.[4] But in his discussion Vivekananda assumes that a "man is great among his fellows in proportion as he can sacrifice for the sake of others." Gandhi or Mother Teresa would have agreed, but not Alexander the Great or Genghis Khan. That is the paradox of humans. We can be either angels or demons, using the very same minds. Not so for other animals.

To bring Wallace's idea up to date, consider that a random selection of babies from any part of the world today could (given adequate early education) give rise to a mixture of all kinds of adult outcomes, such as classical pianists, astrophysicists, plumbers, and artists. This experiment has in fact been performed in modern-day America, the melting pot of all world civilizations, where the children and grandchildren of immigrants go on to do just about everything that any other American can do. But wait, you say: I could tell them apart if I just met them. But you would actually be using visual cues about skin color and so on to make your judgment. You may have heard of the Turing test, which the genius Alan Turing suggested would eventually define the true success of artificial intelligence.[5] The test asks whether you would be able to tell if you were conversing with a computer or a human if you can't see your interlocutor. In other words, could a computer successfully impersonate a human? In fact, no computer has yet passed a Turing test, and there is a "long bet" between Mitchell Kapor (pessimist) and Ray Kurzweil (optimist) about whether one will pass by the year

2029.[6] Let's suggest a modification of the Turing test to determine whether people from different parts of the world are truly different. If you were only allowed to talk on the phone with a third-generation immigrant to the United States (but not see that person), I daresay you would not be able to tell which part of the world his or her grandparents were from—unless you asked directly or questioned that person about world geography.

There are some academic arguments against Wallace's position, but none are conclusive, and it seems equally likely that Wallace was right when he posited that a new evolutionary theory is needed to explain the human mind: "The objections which in this essay I have taken to the view that the same law which appears to have sufficed for the development of animals has been alone the cause of man's superior mental nature will, I have no doubt, be overruled and explained away...they can only be met by the discovery of new facts or new laws, of a nature very different from any yet known to us."[7]

Regardless of the eventual explanations for Wallace's conundrum, most of the human capabilities Wallace was puzzled by would not be possible without our full ToM and multiorder intentionality. We are so much smarter than other animals that our minds seem to be qualitatively different; that is, our brains seem unique in some fundamentally different way. This is why it has been easy for us humans to view ourselves as the designated curators of the world, with all the rest of earth's living and nonliving treasures being placed here for our benefit. In fact, though, our uniqueness is only quantitative. We were able to break through the full ToM barrier by fortuitously hitting upon a solution to the survival-versus-fitness dilemma. Once the crossing of this barrier was accomplished, there was nothing really to hold us back, mentally. Further increases in intelligence, which entailed even more potentially costly mental confusions and obsessions over the complexities of society and daily living, could be accompanied by the continued development of the compensating "executive" brain functions. And, of course, we also were able to question mortality more effectively as we became smarter. So in addition to developing a capacity to filter or ignore reality, we also had to continually enhance our capacity for reality denial.

In contrast to the discrete problem caused by full ToM and advanced awareness of realities such as mortality, the subsequent challenges associated with getting smarter and their continually evolving solutions were likely similar to those faced by any animal subject to positive selection for any useful trait. Thus, although the evolutionary improvements and adjustments took place over generations, once awareness of mortality was "solved" by reality denial, increasing intelligence was likely very rapid on an evolutionary scale. We could become a lot smarter very quickly. This is why we seem qualitatively unique, and why our minds seem so unusual. So we posit that once humans were able to deal with our mortality issues, which stemmed from fully understanding the minds of others, there was nothing to hold us back. Once the shackles were released, we became smart enough to produce more complex language, to orally pass down the knowledge we accumulated over a lifetime, and eventually to write down this knowledge so that we could pass this wisdom on to subsequent generations in an efficient manner. And now, with the advent of the Internet, we can even have a massive and instantaneously updated ToM of billions of humans. Of course, as e-mail, texting, and tweeting have supplanted conversation as a means of communication, this "online ToM" seems to be leaving us (especially the young) much less in real touch with the minds of others. A vast and uncontrolled psychology experiment is under way with the young human minds of today, which will have an unknown impact on society.[8]

As we have already discussed, the ability to innovate, imitate, and culturally transmit information and ideas is a key aspect of human uniqueness. After all, no single individual can invent everything that will be useful to humanity in the future. True innovators are rare, but once innovations emerge, other humans are excellent at imitating them. A striking example is the invention of the number zero.[9] As the late mathematician Tobias Dantzig pointed out, "In the history of culture, the discovery of zero will always stand out as one of the greatest single achievements of the human race."[10] The Greek and Roman civilizations had no clear-cut concept of a zero or a decimal place, making complex mathematics very difficult. And while some other cultures had come close to the concept, religions such as Catholic Christianity even forbade thinking of the concept of nothing, as God was meant to be everywhere.[11] Between the fifth and seventh centuries, mathematicians in India, such as Aryabhata, finally invented and formalized the zero and the decimal place. Some believe that this development occurred in India because no other religious or cultural systems had until then thought of the concept of nothing—we of course take that concept for granted today (the zero appears to have been independently invented by the Maya in the Americas and later lost along with that culture). Regardless of the reasons, a new kind of mathematics developed in India, involving the zero and the decimal place, which was eventually passed on to the Arabs, who applied their Arabic numeral system to it. The Moorish conquests of southern Spain then brought these concepts into Europe, where they spread rapidly. It could be argued that if it were not for these advances in mathematics, many aspects of the European Renaissance might have never occurred. Today, of course, any young child can tell you what a zero is. But it wasn't until around one hundred thousand years after the emergence of the human mind that one individual conceived the idea of a zero. What if that individual had died before conveying the idea to others? What if the group that originally generated the knowledge failed to pass it on? Would we even have a zero today? And would we have modern mathematics and other sciences that depend on this concept?

Cultures and technologies are obviously not static—they also evolve over time, constantly interacting with genes.[12] However, as discussed by Peter Richerson and Robert Boyd in their book *Not by Genes Alone*,[13] the mechanisms of cultural and technological evolution are not the same as those for biological evolution. This important line of thinking (which is related to Richard Dawkins's seminal concept of memes)[14] is further explicated by Alex Mesoudi in his book *Cultural Evolution*, which carries the subtitle *How Darwinian Theory Can Explain Human Culture and Synthesize the Social Sciences*.[15] It is indeed true that one can apply the general principles of natural selection to cultural evolution. But there are also big differences between biological and cultural evolution, as cultural traits can spread very rapidly in time and space. Biological traits are passed down to subsequent generations only when they derive from the molecular structures of the genes, and these changes occur relatively slowly. Natural selection can be highly efficient and produce very dramatic results. But in long-lived organisms like humans, who reproduce infrequently, it takes a long time for natural selection to do its work. This explains why all the various groups of humans who spread out across the planet over the last few tens of thousands of years still have the same general capabilities as those who were left behind in Africa. While subtle variations among various groups or individuals likely exist, the basic aspects of the human mind were already evolved sometime just before the origin of our species, and these core abilities have not changed much since then.

On the other hand, if a member of one generation figures out that water can be had from deep in the ground by digging a well, the next generation knows that easily. The same is true when someone invents a steam engine, or a printing press, or devises a game in which players try to score runs by hitting a ball thrown at them from a distance. The key concepts are innovation and imitation. Newly emerging memes can be incorporated into the technological and cultural heritage of the population promptly. And any useful skill that is developed by one member of the group can spread throughout the population almost instantly. When one human ancestor figured out how to make and sustain a fire, this knowledge could be conferred upon everyone in the group and could also be passed on to other groups. The use of fire to advance human food processing goes back hundreds of thousands of years. The earliest possible evidence is more than 1.5 million years old, and the most definitive evidence is more than eight hundred thousand years old.[16] Thus the invention that now makes it easy to produce and use fire is quite ancient. On the other hand, the ability to produce fire is still not genetically wired in human brains. Humans who have never had any training or knowledge of the different methods of producing a fire would not be able to do so unless they independently invented it again, an unlikely possibility.

In this regard, the Baldwin effect is a much-debated concept that is relevant to the behavioral and cultural evolution of humans. At the beginning of the twentieth century, the Darwinian theory of evolution via natural selection was far from proven. One of the concerns lay in explaining how natural selection could account for very complex behaviors of an unusual nature carried out by some animals. For example, how does evolution select for beavers to be instinctively capable of cooperating to build a dam—a dam that appears to have been intentionally produced to optimally control their own environment? While many papers and books have been written about the Baldwin effect,[17] a very simplified version is as follows. Within a species capable of learning, an individual may have genetic abilities that increase the possibility of learning something new and useful. If that something happens to be beneficial for the whole species, others within the population can pick up the behavior because of their own abilities to learn. Initially they may not be as successful because they do not have the intrinsic genetic ability. But if the behavior creates a niche that is beneficial for the whole group, there will eventually be environmental selection for individuals who are best capable of carrying out the useful behavior. And if this improves the reproductive fitness of those individuals, the genetic ability would *spread more rapidly than it would have by conventional natural selection*. In this manner postnatal learning might bootstrap, or accelerate, the rate at which new behaviors can become part and parcel of the genetically endowed abilities of a species. But it is important to note that the second stage of the Baldwin effect involves genetic "fixation" of the behavior in the species, so that it eventually becomes instinctive.

Critics correctly point out that the Baldwin effect has never been formally proved or demonstrated. And a much more refined and modern concept is Marcus Feldman's concept of niche construction, which applies not only to culture but also to nonlearned behaviors.[18] Anyway, given its relevance to postnatal learning, the Baldwin effect has been suggested more than once as a mechanism to explain the evolution of some of our unique human behaviors and abilities. Indeed, we humans could have experienced a Baldwin effect "on steroids," because of our much superior ability to learn, imitate, and pass on ideas, even beyond the immediate group that observes the original behavior. But, as we said, the first stage of the Baldwin effect is intrinsically unstable because it depends mostly on postnatal learning, a risky thing to rely on for all future environmental circumstances. Thus the second stage of the Baldwin effect, in which the behavior in question eventually becomes genetically hardwired as an instinctive feature of the species, is critical. However, we humans have apparently evolved to not bother with the second stage of the Baldwin effect. While we have invented a great many unusual behaviors (many of them beneficial to our species), essentially all of them continue to require learning by observation and teaching from the previous generation. In other words, we have simply off-loaded the second phase of the Baldwinian mechanism to culture and learning alone, using our many increasingly sophisticated ways of communicating across barriers of time and space. While this would be a risky proposition for other species, we can get away with it, as long as we have a sufficient population size, or a sufficient number of individuals who can remember ideas and pass them on.

And with the advent of writing, computers, and the Internet, we are now largely assured that our accumulated knowledge will be passed on to posterity.

However, this was not the case until a short while ago. For example, archaeological data tell us that when the ancestors of Tasmanians left Africa perhaps fifty or sixty thousand years ago, they carried with them a full suite of basic human inventions, including the ability to make fire. But, as Joseph Henrich summarizes,[19] accumulated evidence indicates that prior to the arrival of European explorers Tasmanian natives had gradually lost numerous valuable skills and technologies, likely including those necessary to make bone tools, cold-weather clothing, hafted tools, nets, various types of spears, and boomerangs. This deterioration appears to have been caused by a combination of a reduction in population size of interacting social learners and the physical isolation that resulted from rising ocean levels at the end of the last ice age (which cut Tasmania off from the rest of Australia for thousands of years). This is a graphic example of the fact that human cultural abilities, unique as they may seem, are largely dependent on imitation and learning and rarely get genetically wired in our DNA—we have to be taught or shown by someone else. Iain Davidson has also eruditely addressed the Tasmanian experience in the context of understanding the origins of human cognition and language.[20]

Here is a thought experiment (gedankenexperiment) that goes further. What if one thousand healthy newborn babies were placed on an island and fully cared for by robots that made no sounds, provided no teaching, and left after the young adults could fend for themselves? What would we find if we came back fifty years later? Over the years, experts from different backgrounds have offered me widely differing answers to this hypothetical query.[21] At one extreme, some question whether these isolated humans would even be standing upright, let alone making any sounds in an organized fashion. Others say that they would have developed some form of sign language (the independent development of new sign languages in communities of deaf children has indeed been documented in Nicaragua[22] and in the Al-Sayyid Bedouin tribe[23]— but these emergences occurred within societies where others were already communicating with spoken language supplemented by gestures). Yet others predict that a new spoken language would have also emerged. Even if the most optimistic scenario were true, these individuals would, mentally and culturally speaking, be back at a point near where the first modern humans were. How long would it then take for them to develop a bow and arrow, or a wheel, or mathematics, or science, or space exploration? Or discover the zero? Perhaps they would never achieve any of these milestones.

In a sense this kind of experiment has been done many times. Let's consider the extreme example of feral children.[24] These are abandoned youngsters who spend the early parts of their lives without any input from other humans. Sometimes they are found living in the wild by themselves or apparently under the care of wild animals, such as wolves. There are many documented cases, including the unfortunate child Genie, who was locked up in a room in Los Angeles from her birth in 1957 until 1970 and never spoken to by her deranged parents.[25] In every case in which such children were discovered after puberty, it was almost impossible to teach them a language, socialize them, or mold them into anything like regular humans. Of course, these sad "experiments of nature" are flawed, because one never knows whether the child in question had mental difficulties to begin with and was abandoned for that reason. And one obviously cannot do a controlled study on normal children to find out.

But experiments on normal children are said to have been done in the past by various monarchs who wished to find out what kind of speech might be produced by children who were isolated from exposure to language from birth. Three such early investigations were supposedly conducted by Psamtik I of Egypt (d. 610 BCE), Frederick II, king of Sicily and later Holy Roman Emperor (1194–1250), and James IV of Scotland (1473–1513). However, authentic and plausible historical evidence for these experiments is quite limited, and if they occurred at all, it is unclear how much can be learned from the limited records.[26] The exception may be the experiment done by the Mogul emperor Akbar the Great (1542–1605), which was recorded in his biography, the *Akbarnāma*.[27] Beyond the fact that it is relatively recent, this story is more plausible, because the *Akbarnāma* is a rare instance in which an intellectually honest monarch apparently asked his biographer to accurately record all information, whether it reflected negatively or positively on him. During an argument about language origins, Akbar is said to have postulated that babies raised without hearing human speech would be unable to speak. To gain experimental proof, the *Akbarnāma* states, he "had a serai built in a place which civilized sounds did not reach. The newly born were put into that place of experience, and honest and active guards were put over them. For a time tongue-tied wet-nurses were admitted there. As they had closed the door of speech, the place was commonly called the Gung Mahal (the dumb-house)." Some time later Akbar was apparently in the vicinity and "he went with a few special attendants to the house of experiment. No cry came from that house of silence, nor was any speech heard there. In spite of their four years they had no part of the talisman of speech, and nothing came out except the noise of the dumb." The likelihood that this experiment did take place is reinforced by other contemporary accounts, such as a letter by Father Jerome Xavier,[28] written in 1598 from Akbar's court: "He [Akbar] told me that nearly twenty years ago he had thirty children shut up before they could speak, and put guards over them so that the nurses might not teach them their language. His object was to see what language they would talk when they grew older, and he resolved to follow the laws and customs of the country whose language was that spoken by the children. But his endeavours were a failure, for none of the children came to speak distinctly." This "palace of silence" experiment is also described in another contemporary account:[29] "An order was issued that several suckling infants should be kept in a secluded place far from habitations, where they should not hear a word spoken. Well-disciplined nurses were to be placed over them, who were to refrain from giving them any instruction in speaking....To carry out this order, about twenty sucklings were taken from their mothers for a consideration in money, and were placed in an empty house, which got the name of Dumb-house. After three or four years the children all came out dumb, excepting some who died there." So it appears that the human mind, like a computer, may be a "nothing in, nothing out" type of mechanism. The basic "operating system" is available, but it needs input and programming. And where does the human mind get most of its input? From other human minds!

Off-loading of important behaviors onto learning and culture (and onto other minds) has also resulted in their evolutionary loss from our genetic endowment. Consider the case of mothering.[30] While all mammals have a maternal instinct, most females (such as female cats and mice) have an instinctive knowledge of what it takes to be a mother and know how to care for their young until they are capable of independent survival. In contrast, Sarah Hrdy points out that the first-born of a female monkey is at some risk, because the mother is not entirely sure what she is supposed to do and needs to learn some things by watching older mothers.[31] The situation is worse with great apes, as a chimpanzee female who has not observed the care of young ones is quite unsure what to do with a new baby and may never become a successful mother.[32] And while basic maternal feelings persist, the details of how to be a successful human mother require education and support from older and more experienced women in the group. Why would natural selection allow something as critical as the genetic wiring for how to be a good mother to get discarded? This is likely because efficient postnatal learning made a genetically determined emergence of fixed brain systems unnecessary.

What all this means is that cultural and technological evolution in humans proceeds at lightning speed compared to the slow plod of biological evolution. In biology, major evolutionary changes that occur over ten thousand years are considered blindingly fast. By contrast, one does not need to be a historian to appreciate the incredible pace of the other evolutionary forces in our lives: Human technology is advancing at a speed that amazes us even in the span of a single lifetime. Automobiles first became available just over a century ago; it is now difficult for most of us to imagine life without them. Powered flight was invented in 1903, and within fifty years it became a common, if somewhat special, form of transportation. Today, commercial air travel is almost like catching a bus, and often less comfortable. We may be part of the last generation who spent more childhood time playing in and experiencing the actual world outside the house than watching it on TV or maneuvering through its representation on a computer screen. This is actually a very important issue, of current concern. Childhood play is a natural phenomenon and likely has a major role in the development of the mind. As we move further and further away from our original ways of growing up, and as we become fixated on TV and computers, it is hard to imagine that children's mental development has not been undergoing significant changes. Child advocacy expert Richard Louv suggests in *Last Child in the Woods*[33] that the lack of nature in the lives of today's children (a "nature-deficit disorder") is the cause of some childhood psychological problems. We are clearly conducting a large-scale social experiment on the young human mind without much knowledge of how this mind originally emerged and what might be the consequences of our experimentation.

Electricity revolutionized our world, making possible many of the useful devices that we use daily. Automobiles revolutionized our world, making it easy to travel great distances with convenience and ease. Television revolutionized our world, allowing us to witness events around the globe as they happen (as well as providing mind-numbing entertainment for the masses). Atomic weapons revolutionized our world, not just with the never-ending threat of annihilation but also by providing some limits on the warfare that can be contemplated by sane leaders (if only humans were generally sane individuals). Commercial jets revolutionized our world, making it much, much smaller. Computers revolutionized our world, allowing us to organize and harness all the other technologies in ways that never could have been imagined just decades ago. And the Internet has now made instantaneous information transfer available to every citizen of the developed world (and even to some in remote areas, without any other modern services). Now imagine a world without most of these innovations. No cars, TVs, planes, computers, Internet, etc. That would be the world of just over one hundred years ago, when we also thought that our Milky Way galaxy comprised the whole universe! It's staggering, really, when you consider how fast things have changed.

Perhaps even more astonishing is the realization that things are changing faster now than ever before. The rate at which we advance technology has increased relentlessly over the course of time. Alvin Toffler's classic 1970 book *Future Shock*[34] was written ahead of its time, but his predictions are now coming true. Toffler pointed out that "too much change in too short a period of time" could create a destabilizing psychological state in individuals and entire societies. The advances in human technologies between 1900 and 2000 were much greater than they were between 1800 and 1900, which were greater than they had been between 1700 and 1800, which were greater than they had been between 1600 and 1700, etc. Yes, there may have been hiccups in brief periods of history, as suggested by the intermittent archaeological evidence that prehistoric humans had remarkable artistic abilities (cave paintings in southern Spain and France; early creative works in Africa tens of thousands of years before that). It is also possible that human mental and cultural development occasionally fell backward, as it did in the dramatic example of the Tasmanians mentioned earlier and as it did during the Dark Ages in Europe, which followed on a period of intellectual ferment in the Greek and Roman Empires.

But the overall rate of technological change has continued

to increase, and there is every reason to believe that this trend will continue. For example, Moore's law correctly predicted that the number of transistors that could be placed inexpensively on an integrated circuit would double approximately every two years (and their cost would go down as well). So as our computer-based devices become smaller and smaller, they are limited more by the size of our fingers than by the electronic components of the device. In a related example, it was only a few years ago that we were overjoyed at the idea of having a telephone that didn't have to be connected to a wire! How amazing this was. Today, if a cell phone cannot also send text messages, take photos, play videos, and connect to the Internet, it is an obsolete toy. And, of course, the phone has to do all this without causing a visible bulge when carried in a pocket and provide GPS capability to boot.

The incredible pace of technology has also created a bemusing new phenomenon. We now see children teaching adults how to use everyday devices. For example, if intelligent, well-educated adults want to know something about new computers or video equipment, most of us just call on one of our kids. It is not uncommon to see a ten-year-old instructing a parent on how to install a new application on his or her hard drive. This sort of thing did not happen much fifty years ago. We accumulated wisdom and talents and passed these on to our offspring. Today, the skills are changing faster than most of us oldsters can keep up with, but the nimble minds of the kids are unfettered by obsolete skill sets and soak up the new technologies like sponges. There was no earlier era in which preteen children were educating adults. The reason is that as cultures advance it is young minds that are capable of absorbing, assimilating, and further developing such advancements. In prior times, advancements occurred at a slow enough pace that this was something only noticed between generations. But now even a couple in their late twenties might find themselves outclassed cognitively in certain areas by their preteen children. It is easy to see that technological change is occurring at a phenomenal pace, and that the rate of change is increasing. However, our biological evolution remains constrained to proceed at a rate that is limited by the rules of genetics. We are still carrying the evolutionary baggage of biological processes that were selected to enhance the fitness of creatures living tens of thousands of years ago. *But most of these cultural advances that have brought us forward would not have been possible without our having acquired a full ToM.*

An additional factor that facilitates technical and cultural advances is that we are each living much longer, which gives us more time to pass on our knowledge to the next generation. But is that just an artifact of altering a physiology originally evolved for an animal with a shorter useful life span? Biblical claims for unrealistically long lives notwithstanding, some data indicate that humans may have already been evolved for relatively long life spans. Despite the common wisdom that Stone Agers did not live beyond twenty-five or thirty-five years of age, Kristen Hawkes and others have pointed out that every present-day hunter-gatherer-like tribe includes a small number of very old people, usually women[35] (we discussed the importance of grandmothers earlier). This contrasts with the life history of captive chimpanzees, who even with the best health care get senile in their forties and fifties, and almost all of them die by age fifty or sixty. So we humans had already evolved to live longer, and this ability is simply now being "unmasked" by the elimination of infectious diseases and other causes of early mortality. Of course, as life expectancy continues to increase, we are carrying out a novel experiment with no past parallel, in which we are artificially prolonging our innate ability to live longer by public health measures and medical intervention. At the present time nobody knows how far this will go. The number of centenarians in some countries has increased by more than 5,000 percent! But as we age we might develop cardiovascular disease or cancer; and if we escape those, then Alzheimer's disease devastates our brains. Our behaviors were also selected for a different type of society, and so we are stressed by the rapid pace of the lives we live—surrounded by strangers, crammed into giant cities, and often relegated to communicating only via text messages.

In addition to our physiological and behavioral processes (as they are classically defined), our biological history demanded that we be masters of denial. Reality denial was essential for us to become smart, and it is now an ingrained component of our biological heritage. But we are not still creatures of denial just because biological evolution is slow. As we became genetically and culturally smarter and smarter, and better able to probe the meaning of reality and mortality using deeper levels of reflection, it was probably necessary for our innate capacity for reality denial to grow as well. Of course this process did not have to be complete, as other mechanisms such as religion came into play for those without adequate culturally enforced denial. Regardless, the point is that this transition has happened only once, despite the fact that smart, warm-blooded social animals have been around for tens of millions of years. Even if biological evolution could now be speeded up, there is no reason to think that we would lose our propensity for denial. Knowing more about the world in general and specifically about the biological workings of living things makes our requirement for denial even greater now than it has ever been. Denial of reality is an essential skill for us to function normally in the world. It is a fundamental property of being human.

Thus we now have a convergence of qualities that make for potentially combustible times. We are an intelligent species that can develop technologies at a blinding pace. But the very intelligence that fuels this technological onslaught demands that we use the massive capacity for reality denial we've been saddled with. We have developed nuclear weapons and built them by the thousands without accepting the inevitability that these will again be unleashed on unsuspecting popu-lations. Moreover, it is increasingly easy to build nuclear bombs (even a desperately poor country like North Korea can do it), and thousands of warheads are supposedly being de-commissioned or simply neglected, especially in some regions of the former Soviet Union. Can any reasonable person really think that this technology will never ever fall into the hands of those who are psychologically less stable than the explo-sive isotopes contained in the weapons? With the advent of global terrorism, we are finally beginning to come to grips with the reality of this possibility. Whether we are being dili-gent enough at this late date remains to be seen.

Similarly, we now know a considerable amount about how new influenza viruses change genetically, enabling them to move from animals to humans and create pandemics that have the potential to kill millions around the world in a mat-ter of weeks. Since at least 1997, we have seen the potential for such events evolving in connection with the Asian bird flu, and experts concur that it is not a question of whether but just when a new killer flu will spread to humans. The bird flu remains mostly confined to birds, and the much-feared major jump has yet to occur as of this writing, although it could any day. Fortunately, it has not happened yet, and there is now work under way to try to develop vaccines or otherwise contain an outbreak. Controversy has also erupted over scientific attempts to genetically engineer a strain capable of more easily being transmitted between ferrets (the experimental animal whose sialic acid targets are most similar to that of humans).[36] However, considering the risks and likely losses, the attention devoted to this matter by our leaders and the media seems inadequate compared to the attention given to more visible and immediate threats. Although both a flu pandemic and terrorism are virtually certain to occur by a rational analysis, the potential bombing of an airplane, for example, is harder to deny and gets more attention than the next flu pandemic, even though losses from the former would be measured in the hundreds and losses from the latter could be measured in the millions. Less obvious but equally dangerous are the sadly popular "vaccine denialists," who incorrectly link these lifesaving procedures to the risk of diseases like autism. Childhood illnesses such as measles and whooping cough are now returning with a vengeance, and unless we come to our senses about vaccines, major epidemics are likely.[37]

Politicians are generally accused of neglecting serious but relatively distant problems, and we typically ascribe this to the relatively short election cycles that drive public policy. If a congressperson has to face reelection in two years, there is little political capital in asking voters to make sacrifices to prevent a possible disaster that might occur in twenty years. This is certainly a factor in our inaction, but there is also a much deeper issue. Unless the catastrophe is actually occurring or is obviously imminent, our innate reality denial mechanisms make it easy to dismiss the issue entirely. It's not a problem; there is no real evidence that it will be a problem; let's not worry about it. Only after planes flew into buildings on 9/11 did citizens of the United States take the terrorist threat seriously. Perhaps only after bird flu is racing through the human population, leaving a trail of carnage without parallel in our history, will we give it the attention it warrants.

There is a foreboding sense that the world is becoming a more dangerous place. Calamities on a massive scale seem to be lurking around the corner. Is this real, or simply a by-product of the information glut that bombards us with horror stories from lands near and far? Unfortunately, it is not an illusion. We evolved to become very smart very fast, in evolutionary time. Our numbers are increasing at an unsustainable rate, and our technology is galloping ahead at an ever-increasing speed. A few of us can destroy cities, countries, and the global environment by pushing a few buttons in a few capitals. We can create a disease pandemic that could spread around the world in a matter of days. We are polluting the earth and changing the climate in ways that we can't predict precisely and likely, at some point, can't easily reverse. If we're so smart, why do we continue to sow the seeds for our eventual destruction? Since we are saddled with a brain designed by selection to cope with the ultimate disaster (death) by denying that it will occur, we also treat other impending disasters by denying that they will ever happen. If we can believe in heaven and reincarnation, how hard is it to convince ourselves that global climate disruption is just an unproven theory? Yes, there is a real danger that our rampaging technology will come back to bite us in a big way, and very soon. But is this inevitable? Maybe yes, but maybe no, if we can recognize and mitigate our penchant for denial.

CHAPTER 10

A Tale of Two Futures: Are You a Pessimist or an Optimist?

We're back in denial. If people are saying there isn't a problem, that's part of the problem.
> —Mary Howell, MD, cofounder of the National
> Women's Health Network

Your efforts must be realistic. We must not rely on appearances. We must do research to determine what is real. With no research, you can't find reality.
> —Tenzin Gyatso, the Dalai Lama, in a speech at
> UC San Diego[1]

The potent combination of our powerful intelligence with our massive reality denial has led to a dangerous world, some examples of which we have mentioned. Less obvious, but in the long term more dangerous, are threats resulting directly or indirectly from technological developments that have permitted us to increase our numbers well beyond the carrying capacity of the natural world. More efficient agriculture and the invention of artificial fertilizers permitted humans to produce food sufficient to support numbers that would be unthinkable for other animals of our physical size. Public health measures, vaccinations, antibiotics, and other medical advances also permitted population numbers to explode. The world is overpopulated already and is becoming more so at an alarming rate. And although we pay lip service to the resulting problems, we do relatively little to address their root causes. Indeed, some religions continue to promote the unrestrained propagation of their flocks. Planet Earth is sick, with a bad case of "infection by humans." In fact, as far as the other species on the planet are concerned, we humans are like the rapaciously invasive conglomerate of aliens called the Borg in the classic TV series *Star Trek*—a race that indiscriminately assimilates and takes over anything and everything it encounters. The motto of the Borg is "Resistance is futile." And indeed, for all other species on Planet Earth, resistance is futile when faced with humans! The exceptions, of course, are the microbes that infect us (such as tuberculosis, HIV, and malaria), which are also spreading just fine, thank you. As explained by environmental activist Paul Gilding in *The Great Disruption*:[2] "We have now reached a moment where four words—the earth is full—will define our times. This is not a philosophical statement; this is just science based in physics, chemistry and biology....To keep operating at our current level, we need 50 percent more Earth than we've got."

The most dramatic consequence, of course, is the effect we are having on the atmosphere and the climate. Besides the very public efforts of Al Gore,[3] many writers have spoken out about this vital issue, including *Scientific American* editor Fred Guterl. *The Fate of the Species: Why the Human Race May Cause Its Own Extinction and How We Can Stop It*[4] describes climate change as one of the most pressing dangers to the human species. But while this has become a popular topic of discussion, few are willing to make the major lifestyle changes necessary to reverse it. The government of the country that is one of the biggest per-capita culprits (the United States) now at least acknowledges the reality of global climate change—but still refuses to face up to the problem. The energy platforms of major political candidates for leadership positions in the United States carefully ignore or minimize attention to this politically charged issue. And the same is true of the other major contributors to the problem, such as those companies that extract more and more fossil fuels from the earth yet run misleading advertising campaigns that claim that they really do care about the environment. Even worse, the melting of Arctic ice has given many countries the impetus to prospect for more fossil fuels in that pristine wilderness.

Why is it that ordinary citizens do not sit up and take notice of the danger? Unfortunately, the focus remains mostly on "global warming" instead of on the bigger concern—that we are disrupting the planet's climate in completely unpredictable ways. Because climate prediction includes a significant degree of scientific uncertainty, this has allowed skeptics to gain the upper hand and even corner some expert scientists into difficult positions. A friend in the climate research field privately admits that he and most of his colleagues are afraid to stand up and speak out because of the vituperative attacks and massive smear campaigns that they would inevitably suffer—as did Michael Mann[5] and others. But much research indicates that as forests disappear and polar ice caps melt, etc., there are unpredictable feedback mechanisms that will make global warming increasingly difficult to tackle. Even more worrisome, there will likely be a tipping point after which continued warming may become irreversible, no matter what we do. Of course, other scenarios are also possible. For example, it is plausible that we could instead tip the planet into an ice age. The Hollywood movie *The Day after Tomorrow* took a reasonably valid climate change model that is possible over a sixty-year span and instead told a story in which it happened in six weeks. This made it easy for viewers to deny the possibility that this could ever actually happen.

During Bill Clinton's successful bid for the US presidency, one of his campaign aphorisms was "It's the economy, stupid!" The point was that while strategists were considering many diverse and important political issues, the state of the economy was the single factor that was actually going to determine the outcome of the election. In like manner, we humans are focusing on the wrong issues when it comes to the debate over global warming. The slogan should be "It's local climate destabilization, stupid!" One does not need to be an expert to find convincing evidence that global temperatures are indeed rising, and that the climate is changing, likely due to human activities. Every one of the more than 150 national scientific academies in the world, every professional scientific society with members in relevant fields, and more than 98 percent of all scientists who study climate agree on this point. There is increasing agreement that the Industrial Revolution ushered in a new climatic period, which Nobel Prize winner Paul Crutzen has called the Anthropocene.[6] The frequency of extreme weather events across the world is increasing at a rate not previously seen since climate records began to be kept[7]—and an ice-free Arctic sea may be years, not decades, away.[8] However, despite the overwhelming body of data, it is unclear exactly what is going to happen in the future. Thus the mild-sounding term *warming* is too easy to pass off as being irrelevant to an individual ("So what? I will just turn up my air conditioner!").

Instead of allowing complacency based on uncertainty, we need to look back at the history of climate on this planet and consider the potential consequences of human interference. Data from sources such as the Greenland ice core (from which it is possible to determine historical temperatures over relatively short time spans) indicate that the period until around ten thousand years ago was prone to wildly oscillating temperatures (similar data is available back to almost a million years ago).[9] For example, twelve thousand years ago there were likely tens of feet of ice over what is now San Francisco and Washington, D.C., as well as over most of northern Europe. The less frequent and cyclical warm periods of the past could also be associated with temperature fluctuations, variations in ocean levels, and so on. In contrast, the last ten thousand or so years have been one of the uncommon periods of *relatively stable warm climate*. Years ago I used to joke that this must be a consequence of humans having spread all across the planet at approximately that time, reaching all the way into South America and most other habitable parts of the world. In fact, this concept is now the basis of a current theory[10]—that the diaspora of humans, and the associated burning of forests, the initiation of agriculture, and the elimination of most large animal species (beginning around ten thousand years ago) may have stabilized the climate and prevented its usual wanderings. Whatever the mechanism, we are living in an unusually stable period called the Holocene epoch. It is only because of this stable climate that we have so successfully populated the world while optimizing our homes, facilities, and agriculture *to suit a relatively predictable local climate in each location.*

So what the average human should fear is not global warming but rather local climate destabilization, i.e., a change in the relative stability and predictability of his or her own local situation. This does not only mean unusually severe hurricanes and tornadoes, or unexpected droughts and flooding. Such sad catastrophes affect only a small part of the world at any one time. Of even greater concern should be local changes that may seem trivial yet have a huge impact on our living conditions and economies. Try a casual poll of your friends across the planet who have lived in the same place for a while, and ask, "How's the weather been lately in your neck of the woods?" There is a high probability that the answer will include words such as "strange," "unusual," and "weird." And in some cases your friends will not mention warming but rather unusual cold spells, or rain and snow that fell with unusual frequency at unexpected times. The general trend seems to be increasing dryness in previously dry areas and increasing wetness in previously wet areas. It will not take many more such changes to disrupt local economies and agriculture in a manner that destabilizes local societies. And the impact of local events can be global. For example, very high temperatures in Russia in 2010 were unpleasant, but the bigger consequences were forest fires and loss of wheat production. The unprecedented 2011 floods in Thailand raised the costs of computer hard disks worldwide because some key local factories were damaged. And in 2012, the great drought in North America decimated the corn crop and ignited forest fires that destroyed many homes.

Remarkably, the subject of climate change was never brought up by the moderators of the three US presidential debates of 2012—and was only mentioned briefly by Barack Obama in his acceptance speech upon reelection (he did expand on the theme in his 2013 State of the Union address). It remained the big elephant in the room that everyone was conveniently ignoring. But as this is being written, the northeastern seaboard of the United States is still struggling to recover from the devastation wrought by the deadly hurricane Sandy. This so-called Frankenstorm was unprecedented in terms of its timing, path, and site of landfall. The hurricane was apparently unique in the annals of American weather history, and many climate scientists feel it is the latest manifestation of human-induced climate change. It is particularly ironic, then, that Sandy is a diminutive for Cassandra, the mythical and tragic daughter of the king of Troy. The Greek god Apollo was so smitten by Cassandra's beauty that he gave her the gift of prophecy, which enabled her to accurately predict the future. But Apollo became angry when Cassandra spurned his amorous advances. So he placed a curse on her, assuring that nobody would believe her prophecies. Cassandra finally became insane because her predictions of future events fell on deaf ears and she was powerless to prevent catastrophes she knew would occur. The Cassandra metaphor has been applied to many different domains of human activities, and we can now add climate change to the list. Despite being fully aware of the likelihood of serious climate disruption and its potentially deadly consequences, climate scientists seem to be almost powerless to convince the public at large of the danger.

But the tragedy of Hurricane Sandy may have a silver lining. Politicians whose constituents were directly affected began speaking out. The New York State governor, Andrew Cuomo, said in a news briefing the day after the hurricane hit: "There has been a series of extreme weather incidents. That is not a political statement. That is a factual statement. . . .Anyone who says there's not a dramatic change in weather patterns, I think, is denying reality." The next day he added, "I think part of learning from this is realizing that climate change is a reality." And the politically independent New York City mayor, Michael Bloomberg, said: "Our climate is changing. And while the increase in extreme weather we have experienced in New York City and around the world may or may not be the result of it, the risk that it may be— given the devastation it is wreaking—should be enough to compel all elected leaders to take immediate action." Fittingly, the contemporaneous cover story of *Bloomberg Businessweek* magazine was entitled "It's Global Warming, Stupid."

While the public and politicians may choose to ignore climate change, the insurance industry cannot afford to do so. Using its comprehensive NatCatSERVICE database, which maintains data on natural catastrophes, Munich Re (one of the world's largest reinsurance conglomerates) analyzes the frequency and loss trends of various events from an insurance perspective. In an October 2012 report, Munich Re published its analysis of all kinds of weather perils and trends. The study was prepared for underwriters and clients in North America, the world's largest insurance and reinsurance market. Ironically, this region (also one of the world's largest producers of greenhouse gases) has been most affected by extreme weather-related events in recent decades. The North American continent is already vulnerable to all types of weather hazards—hurricanes, winter storms, tornadoes, wildfires, drought, and floods (one reason is that there is no mountain range running from east to west that can separate hot from cold air). But the study shows a nearly fivefold increase in the number of weather-related loss events in North America for the period from 1980 to 2011, compared to a 400 percent increase in Asia, a 250 percent increase in Africa, a 200 percent increase in Europe, and a 150 percent increase in South America. The overall loss burden during this time frame from weather catastrophes in the United States was $1.06 trillion (in 2011 values), and some thirty thousand people lost their lives.

Just a few weeks before the Munich Re report appeared, James Hansen and other scientists at NASA's Goddard Institute for Space Studies, in New York, published a study[11] on the apparent increase in extreme heat waves during the summertime. Such events, which just a few decades ago affected less than 1 percent of the earth's surface, "now typically cover about 10 percent of the land area," the paper stated. "It follows that we can state, with a high degree of confidence, that extreme anomalies"—i.e., heat waves—"such as those in Texas and Oklahoma in 2011 and Moscow in 2010 were a consequence of global warming because their likelihood in the absence of global warming was exceedingly small."

Meanwhile, in a Kafkaesque twist, the late-2012 United Nations Framework Convention on Climate Change took place in Qatar, the country that tops all others in per-capita production of greenhouse gases! The official meeting report sounded superficially encouraging:[12] "Countries... agreed on a firm timetable to adopt a universal climate agreement by 2015....They further *agreed on a path to raise the necessary ambition* [italics mine] to respond to climate change, endorsed the completion of new institutions, and agreed on ways to deliver scaled-up climate finance and technology to developing countries. The Kyoto Protocol...under which developed countries commit to cutting greenhouse gases...will continue...the length of the second commitment period will be eight years." But Greenpeace International responded:

"Which planet are you on? Clearly not the planet where people are dying from storms, floods, and droughts....The talks in Doha...failed to live up to even the historically low expectations. Where is the urgency?....It appears governments are putting national short-term interest ahead of long-term global survival."[13]

Conservation International concluded: "There has been virtually no meaningful progress on any important issue.... At most, what this meeting has achieved is an agreement to continue negotiating next year. This is completely unacceptable and irresponsible....They are playing a game of chicken that risks sending all of us over the cliff."[14]

Even more worrisome, it was hard to find any headline or front-page news about the all-important Doha talks.

Indeed, the mainstream media seem to have generally lost interest in this story, perhaps because it has become so commonplace. This despite the fact that the National Oceanic and Atmospheric Administration (NOAA) has just officially declared that 2012 was the hottest year on record for the continental United States and the second-worst for "extreme" weather conditions, such as hurricanes, droughts, and floods.[15]

Seven of the ten hottest years in US records (dating back to 1895), and four of the hottest five, have followed 1990, according to NOAA figures. The year 2012 also saw Arctic sea ice hit a record low, based on more than thirty years of satellite observations. And at a global level, according to NOAA scientists, all twelve years of the twenty-first century so far (2001–2012) rank among the fourteen warmest in the 133-year period since records have been kept.[16]

Despite all the facts, there is a deafening silence on the part of world leadership over the plausible relationship between extreme weather events and human-induced climate disruption. The analogy that comes to mind is the emperor Nero playing the fiddle while his city, Rome, burned down! But it is too late now, anyway—the "Planet Earth Climate Destabilization Experiment," as I call it, is under way, and we just have to wait to see how bad things are going to get. We have poked a tiger in the eye, and we just have to hope we will get away without suffering too many consequences. We should be very concerned about disrupting the relatively predictable weather that we currently enjoy, something that is critical not only for human living conditions in most parts of the world but also for the relative stability of our local economies and livelihoods. Nevertheless, it is more convenient to simply deny the problem. Indeed, the news media seem to have gotten weary of reporting extreme weather events and don't bother anymore unless the events are close to home. Were readers in America aware, for example, that Pakistan had a *second* megaflood in 2011? Both of these catastrophic floods affected a single area, washed away vital crops, forced almost two million people to flee their homes, and left them suffering from malaria, hepatitis, and other diseases. And did you hear that the city of Beijing had record-breaking floods in 2012? The heaviest rainfall to hit China's capital in sixty years left many dead, stranded thousands at the main airport, and flooded major roads. Almost two million people were affected, and economic losses were estimated at $1.5 billion. But this kind of event is no longer considered international news—as it has become far too common.

In the worst-case scenario, we could even find ourselves enduring the same wild weather that plagued times past, swinging between ice ages and warm periods. But the skeptic can argue that current climate models might turn out to be wrong after all, and that climate destabilization might not be as severe as the alarmists suggest. And as Matt Ridley points out in *The Rational Optimist*, some prior doomsday predictions turned out to be overstated. For example, Rachel Carson's 1962 book, *Silent Spring*,[17] documented the detrimental effects of pesticides on the environment, particularly on birds. The title was meant to evoke a future spring season in which no bird could be heard singing because they had all vanished due to pesticide abuse. Given the state of knowledge at the time, Carson's additional concerns about effects of synthetic chemicals on human health were reasonable, and many have been borne out. But decades later some suggest that the overreaction to this seminal book also did some harm—for example, by eliminating the vital role that DDT played in killing malaria-carrying mosquitoes. But it also seems that the resulting cleanup of toxic lead and mercury from the environment has had a positive effect on the brain development of children. Another example that Ridley cites is Paul Ehrlich's 1969 book, *The Population Bomb*,[18] whose original edition began with the statement, "The battle to feed all of humanity is over...nothing can prevent a substantial increase in the world death rate." Ehrlich and his wife, Anne, still stand by the basic ideas in the book, believing that it achieved their goals because "it alerted people to the importance of environmental issues and brought human numbers into the debate on the human future."[19] And there is no doubt that many issues related to population growth were addressed only because of these warnings. But the book also made a number of specific predictions that did not come to pass. Ridley argues that, guided by our human ingenuity and ability to adjust to change, we took action to fix some of these problems, and the predictions turned out to be worse than the reality. So his comforting notion is that humanity can and will fix the climate problem when the time comes to really deal with it.

But there is one *big* difference between pesticide pollution and population growth on the one hand and global climate change on the other. Unlike *any other current policy issue*, the potential for climate destabilization is one we *simply cannot afford to get wrong the first time around*. The fact is that we humans are conducting a very dangerous experiment with our climate on *the one and only* planet that we have. Whether or not the recent spate of extreme weather events is a harbinger of things to come, the point is that there is *no way of turning back* once we have set major climate destabilization in motion. And if it happens, the consequences will be devastating for all of humanity at both the local and the global level. So unlike almost any other policy issue, about which we can afford to debate the pros and cons and change our minds later as new information comes in, *there is no margin for error here*. There is only *one* planet, *one* biosphere, and *one* Anthropocene epoch, and we must err on the side of caution!

Think back to the time of the Cold War and the "mutually assured destruction" that was guaranteed if nuclear war broke out. Everyone agreed that once that genie was out of the bottle, there would be no way to go from nuclear winter back to the status quo. So extreme caution was the name of the game, and numerous fail-safe mechanisms were put in place to avoid nuclear war *at all costs*. The same is true of the potential for irreversible global climate destabilization. Once we have tipped the balance, we simply do not know how to reliably turn the clock back. And if we are eventually forced to institute the many geoengineering (planetary climate intervention) approaches currently being considered,[20] we face even more unknown risks, including the possibility of making climate destabilization worse rather than better.

So even if there is a *10 percent probability of human-induced climate destabilization*, we should take it very seriously. And rather than haggling over whether or not global warming is occurring and how serious it is, we should all insist on a program of research geared toward global climate maintenance, so that we can understand why the climate has been so relatively stable for the last ten thousand years and do whatever is necessary to *maintain this stability*. This approach might bring people together for a common goal, and the changes needed may not be as draconian as suggested by extreme environmentalists or as trivial as those suggested by growth-oriented businesspeople. One fervently hopes that humans will not continue to deny the reality of this risk at our peril. To repeat, humanity should be realizing that *unlike any other problem that we face*, the outcome of climate destabilization, whatever it is, *would be irreversible*. In other words, *we only have one planet*, and it does not matter if it turns out that climate-destabilization concerns are overblown after all. The point is that if the concerns are even partly real, there will be global climate disruption, which at the very least will be extremely bad for the economies of many parts of the world. And once the climate is "broken," it will be like Humpty Dumpty—impossible to put back together again.

Many approaches have been taken to try to convince individual citizens to take climate change seriously. These include the arguments that we have an ethical and moral obligation to the less fortunate on the planet and to future generations,[21] that we are conducting a risky experiment with the planet, that the global economy will suffer, and so on. However, none of these really has had an effect because of our all-powerful capacity for reality denial. The resulting optimism bias makes most people (even if they believe the science) just go on with life and hope for the best. Between the time the writing of this book began, in 2007, and the time it was completed, toward the end of 2012, much change has occurred, and there has been a spate of extreme weather events. Regrettable as they are, these events have at least made many people begin to think that climate change may be real. However, as long as there is no certainty regarding the future of climate, the average person will continue to hope for the best, and the naysayers and profiteers will take advantage of that uncertainty.

Even appeals to individuals based on the fact that local climate disruption will affect their lifestyles and pocketbooks generate the response "How can you be certain?" So let me offer an analogy that the average human can perhaps better understand and relate to. Imagine you are going to take a long airplane flight and you're told that there is just a 10 percent chance that things will go badly wrong and that the plane will crash. Would you get on that plane? Of course not: Most people would want at least a 99 percent certainty that the flight is safe. Well, the same is true for climate disruption. Even if the probability of extreme climate disruption and major irreversible damage to the planet is only 10 percent, we should treat it as if it were an airplane that has a 10 percent chance of going down—with deadly seriousness.

By the time you're reading this, a majority of skeptics may have finally begun to admit that something bad is happening to our climate and that we humans may be contributing to it. But given the extreme degree of polarization surrounding the debate, it is unlikely that any consensus on action will be reached soon. If so, there are alternate sensible approaches that can be pursued. Durwood Zaelke, founder of the Institute for Governance and Sustainable Development, and Veerabhadran Ramanathan, a professor at the Scripps Institution of Oceanography, suggest a short-term strategy that involves cutting emissions of four climate pollutants: "black carbon, a component of soot; methane, the main component of natural gas; lower-level ozone, a main ingredient of urban smog; and hydrofluorocarbons, or HFCs, which are used as coolants. They account for as much as 40 percent of current warming. Unlike carbon dioxide, these pollutants are short-lived in the atmosphere. If we stop emitting them, they will disappear in a matter of weeks to a few decades. We have technologies to do this, and, in many cases, laws and institutions to support these cuts."[22] Another reasonable approach is suggested by Peter Byck in his movie *Carbon Nation*.[23] Byck points out that even if you agree to disagree with those who deny climate disruption by humans, most such denialists are still very much in favor of clean air, clean water, and clean sources of energy. One can even point out that there are opportunities to make good money in connection with several of the new approaches to alternative energy. Perhaps this is the way to bypass our human denial of climate change and deal with the problem. In yet another approach, perhaps popular writers may succeed where scientists have failed. In her novel *Flight Behavior*, Barbara Kingsolver[24] uses fiction as a device to draw attention to the disparate human reactions to climate change. In the novel, when an unexpected but spectacular swarm of monarch butterflies appears in Appalachia, a phenomenon usually seen in Mexico, it is interpreted by scientists as clear evidence of climate change—but is instead seen by devout local Christians as a miraculous sign from God.

Again, let us hope it is not too late to turn back the clock so that we can continue to have a relatively stable climate, as we have enjoyed in the last ten thousand years. But as of this writing, we simply continue to deny the limits to which we can push the planet.[25] And Paul Ehrlich is right that there are still potential disasters tied to overpopulation. We are programmed by our evolutionary drive to reproduce. The problem of overpopulation has gone from being well recognized to being almost denied in some countries. The consequences are not simply overconsumption of resources but also acceleration of global climate disruption. But as female literacy increases, population growth stops and even declines.[26] Not surprisingly, it seems that as soon as women get educated, they don't want to be baby-making machines anymore. So if *female literacy and independence can spread*, population growth may come under control. Regardless, even those of us who understand climate disruption and population problems deny them on a day-to-day basis or simply don't do much about them.

We are also using the earth's natural resources at an alarming rate, and they will become more scarce and costly. Forests are disappearing, especially in the tropics. Clean water is becoming increasingly difficult to secure. Dwindling resources were already a problem when only a small part of the world's people were living in developed countries. Now, as the huge populations of other parts of the world aspire to similar levels of resource-gobbling lifestyles (especially in eastern and southern Asia), the rate of utilization of limited natural resources will increase, as will the rate at which we produce pollutants that poison the globe. And, of course, as resources become more limited, the chances become greater that those nasty nuclear and chemical weapons will be used in anger.

Increased population density and rapid global transportation also increase the potential for danger from biological pestilence. Even the spread of HIV (which is not especially contagious) has not been effectively contained. In addition, our technology makes it possible for us to construct horrific biological weapons with relative ease. These types of weapons have often been considered impractical because one cannot easily confine the devastating effects to the target population. For an insane terrorist, though, confinement may not be much of a concern, and a properly designed infectious agent could travel the globe in a very short time. We mentioned the controversy over a laboratory version of the bird flu virus that was engineered to spread rapidly in humans. While the scientists who created this dangerous virus did so with the sincere intent of finding ways to protect us against it, there is indeed a small risk that a terrorist group could discover and misuse this knowledge. Of course, this assumes the terrorists are not intelligent enough to realize that such a virus could just as well circle back to hit their own societies even more violently than it would hit the first victims.

One can build a logical argument that our innate reality denial, coupled with our runaway technological achievements, virtually guarantees that we will be facing global calamities on a scale never before seen. Many different scenarios can be constructed around resource depletion, climate change, disease pandemics, etc., that will lead to a breakdown in modern civilization, war, and human death and suffering on a massive scale. History suggests that we will not learn any long-term lessons from the first few of these disasters, in large part because of our nemesis, reality denial. Indeed, it is arguable that we are destined ultimately to destroy ourselves as a species—or, at the very least, to continue to cycle between well-developed civilization and catastrophic collapse, never reaching a technological state much beyond what we currently enjoy. We hope that these words do not prove to be prophetic. But we may well be in for a cycle of catastrophic collapses and have to rebuild ourselves, much as many civilizations have done in the past.

Interestingly, the fundamental forces that got us to this dilemma *might be valid for any planet that harbors intelligent life*. The rules of natural selection, and the fitness-versus-survival paradox that arises when attaining a full ToM (or the alien equivalent), do not depend on what types of molecules make up the life form in question. The generality of deep-seated self-deception (denial) as the primary solution to the full ToM problem of mortality awareness may in fact be the only possible universally viable mechanism. In other words, if this pessimistic view of our future is true, any intelligence at the humanlike level may be in a metastable state, wherever it evolves. This is one of the potential solutions to what is referred to as the Fermi paradox—Enrico Fermi's question about why, if there are so many planets in the universe (and even in our galaxy) that could harbor life, have we never seen hard evidence that our neighbors exist or have visited us? In other words, "Where is everybody?"[27] It's possible that the solution to the full ToM barrier (the requirement for reality denial) ensures that no civilization of intelligent species will persist long enough to master long-distance space travel, either manned or unmanned. SETI projects (those concerned with the search for extraterrestrial intelligence) have so far failed to find intelligent life elsewhere in the cosmos. The only way to have the kind of intelligent life we can communicate with is for other beings to also have full ToM (or the equivalent). And if that requires denial of reality, then such other civilizations (if they do exist) are in a similar situation, trapped in the same cycle of denial and self-destruction that we describe.

Some years ago I was invited to speak at a conference called Bioastronomy: Life among the Stars, at which I became aware of the Drake equation,[28] which makes an estimate of the probability that intelligent life exists somewhere else in the universe. In 1961, Frank Drake of the University of California at Santa Cruz created this equation during his preparations for the first meeting about detecting extraterrestrial intelligence. The equation tries to predict the number of civilizations in our galaxy with which communication might be possible, taking into account such factors as the average rate of star formation per year in our galaxy, the number of those stars that have planets, the average number of such planets that can potentially support life, the number of those planets that go on to develop life at some point, the number of those planets that actually go on to develop intelligent life, the percentage of those intelligent civilizations that develop a technology that releases detectable signs of their existence into space, and the length of time for which such civilizations release detectable signals.

Since 1961, the known number of stars with planets and the percentage of those that could potentially support life have increased markedly (especially since the Kepler telescope was launched in 2009). On the other hand, many scientists (including the great biologist Ernst Mayr) have already pointed out that of the billions of species that have existed on earth, only one has become intelligent enough to ask the question. So the number of life-sustaining planets in which civilizations might develop a technology that releases detectable signs of their existence into space must be very, very small. Now, if our theory is correct, it further drastically alters the calculation—because one first has to develop a life form with full ToM (or the alien equivalent) that can communicate with others of its kind before trying to communicate with intelligent life forms on other planets. And we can suggest that this is an extremely unlikely probability, just as it was on Planet Earth. Furthermore, reality denial will likely be part and parcel of any such civilization, hastening its demise. So it might well be that we are all alone in our galaxy and even in the known universe, asking ourselves who we are and how we got here.

Even if this proves to be wrong, most of us would like to believe that our alien visitors would follow the *Star Trek* Prime Directive—noninterference in the internal development of other civilizations.[29] In fact, though, a policy of noninterference is unlikely to be followed by any civilization. Throughout earth's history, whenever two civilizations have met, the less advanced one usually ends up being destroyed. And since humans have only recently thought about looking for extraterrestrial life, chances are high that the other life forms are more advanced than we are. So if we do meet them, it is likely to mean the end of our civilization. Thus it could be a very bad idea for us to try to find an alien civilization. As someone at the bioastronomy conference exclaimed, "If the phone rings, don't answer!"

Let's now come back down to earth—pun intended. Even those of us who agree that human nature and technology are essentially incompatible would like to think that eventually, perhaps after a disaster or two, we would shape up and come to grips with our basic problems. One can be an optimist in this manner and still accept all the arguments here about self-awareness and denial. Indeed, it is probably *essential* for our long-term success that we embrace the idea that reality denial is a fundamental part of human nature. For only by knowing this enemy can we consciously change our innate, destructive behavioral tendencies. Ironically, many readers of this statement are likely to deny the important point we are making. In other words, *we are in a state of denial about our denial of reality*, and this is not an easy problem to overcome on a daily basis.

It is only by understanding reality denial as an enemy within that we might be able to overcome it. An alcoholic is not necessarily a drunk. But in order to avoid becoming one, it is necessary for the alcoholic to acknowledge his innate tendencies and to actively fight the impulses that drive a behavior that is satisfying in the short term but self-destructive over the long haul. Just as an alcoholic must sometimes hit psychological or emotional bottom in order to come to grips with his problem, it may require a small nuclear war or a major climatic disruption in order for us to see the light. We haven't seen it yet, nor have we even acknowledged the underlying trait (denial) that makes it so difficult for us to deal with our critical issues. As Robert Trivers puts it: "Denial is also self-reinforcing—once you make the first denial, you tend to commit to it: you will deny, deny the denial, deny that and so on."[30]

Many of our everyday problems also have a component of denial—whether we're investing in a risky scheme, deciding to stay with an abusive partner, or whatever. Those who give advice professionally or otherwise can easily detect the denial component in a person's situation and advise the subject to escape from the danger. What they may fail to recognize is that reality denial is such a fundamental part of being human that one cannot easily just shed its clutches. To continue the alcoholic analogy, we can't abandon denial any more than we can change our fundamental personality traits. What we can do is to recognize this trait and try to manage its deleterious effects, just as the alcoholic manages his disease. Indeed, we don't want to completely escape from our state of denial, even if we could—it's the only thing that keeps us sane in the face of rational realization of mortality. We just need to recognize and manage its pathological consequences. We need psychotherapy on a societal and global scale.

The big question is, then, are we capable of controlling denial sufficiently to solve our current dilemmas? Can we create a spiritual construct (individually, or as a new formal religion) that can satisfy our acceptance of mortality without letting it drive our lives and society to oblivion? As always, the first step is recognition of the problem. The next step may require a process that is every bit as unlikely as the convergence of self-awareness and self-deception that allowed us to break through the wall so many years ago. It required millions of years for the latter event to occur. We don't have that much time to solve our current dilemma. Humans may be products of chance events that allowed full ToM and intentionality to emerge, but we were able to come into existence only because we simultaneously developed the ability to deny our own mortality and reality. But as a by-product, we also deny many of our other problems, despite having the ability to understand them.

The twentieth century was the era in which the greatest amount of human knowledge was accumulated, knowledge that was contributed to by people who were following up on the knowledge generated by many past human civilizations. This knowledge base (which is essentially the understanding and appreciation of aspects of reality) generated much scientific and technological progress, and there was great hope that the twenty-first century would see further steps in this direction. However, a variety of sociopolitical doctrines and agendas seemed to have caused a marked regression in the public appreciation for the value of scientific knowledge.

In 2007 I had the opportunity to participate in a historic meeting in which some of the oldest and best-known American societies that focus on the value of knowledge (the American Philosophical Society, the American Academy of Arts and Sciences, and the National Academy of Sciences) came together to discuss the situation in America. This meeting was entitled The Public Good: Knowledge as the Foundation for a Democratic Society and involved scientists, humanists, and leaders in business and public affairs from around the country in a two-and-a-half-day series of panel discussions, conversations, and dinner programs that focused on "some of the most pressing issues facing the nation." The underlying premise was simple—that in order for a democracy to succeed it should be based on real knowledge of the facts of the world around us. Following the meeting, several tomes were published to explicate this important though obvious idea.[31] But this effort has been largely ignored, and reality denial has continued to gain ground. Why is this so? Perhaps it is because reality is unpleasant for most people, and our built-in mechanisms for denial allow us to pick up whatever line of reasoning we find the most comforting in the face of competitive realities, however flawed that reasoning is. But let us hope that other efforts like this continue, so that we can once again base our future on our understanding of reality.

Sadly, following a century of intense focus on the value of science for society, we are even facing a growing and dangerous antiscience movement that appears to originate from adherence to a variety of social, political, and religious doctrines that favor alternate realities. *New Yorker* staff writer Michael Specter addresses this new and widespread fear of science and the consequences of this reality denial for individuals and for the planet in his 2009 book, *Denialism*.[32] He expresses concern over the fact that both political leaders and the public seem to mistrust science more than ever before. So irrational and unfounded fears about everything from childhood vaccines to genetically modified grains abound, even while dietary supplements and "natural" cures with no proven value are gaining many followers. As Specter sees it, this war against science amounts to a war on progress itself, and it's occurring at a time when we actually need science more than ever to chart our future in a rational fashion.

Why is it that so many humans are attracted to these illogical doctrines? Is it simply a reflection of the fact that most humans would prefer not to face reality? After all, science is all about revealing factual realities. But is it possible to hold a full ToM along with a complete recognition of reality? Yes, of course it is, but it requires a large amount of definitive information and rational thinking about the meaning of all that information. We mentioned earlier the excellent book by Richard Dawkins entitled *The Magic of Reality*. In this wonderfully written and beautifully illustrated volume, Dawkins explicates the reality of who we are, starting with the Big Bang, proceeding through human evolution, and going all the way to atoms and subatomic structures. As Dawkins correctly points out, this true reality (as revealed by the methods of science) is indeed magical beyond belief, at least from our perspective as humans. However, it is easy for someone of Dawkins's knowledge and intellect to appreciate reality's magic. For the average human, though, the enormity of this reality can actually be rather unpleasant, and the book could also be read as *The Horror of Reality*. Thus most people prefer to ignore or rationalize away many of the realities they do not like to think about. On the other hand, even the extreme realist cannot get through a day without ignoring some realities and taking some risks. The ideal situation, therefore, would be to relish and use both our full ToM as well as our ability to deny reality, allowing us optimism. And indeed, as we shall see, there are also many benefits to reality denial.

CHAPTER 11

On the Positive Value of Human Reality Denial

> Hope is the denial of reality.
> —Margaret Weis, in *Dragons of Autumn Twilight*

> My uniform experience has convinced me that there is no other God than Truth.
> —Mohandas Karamchand Gandhi, in *An Autobiography*

> Whatsoever thy hand findeth to do, do it with thy might; for there is no work, nor device, nor knowledge, nor wisdom, in the grave, whither thou goest.
> —Ecclesiastes 9:10

The focus of the last chapter was on the negative consequences of the human penchant for denial of reality, with only a few suggestions for the optimist. Let's now discuss the fact that both denial and recognition of reality can have powerful positive effects on personal, social, communal, national, and global matters. In other words, there are many valuable aspects to this unusual human capacity, especially when combined with our ability to identify with the minds of others.

Let's begin with a topic I am personally familiar with. Having practiced for some years in the past as a part-time clinical oncologist, I can say that reality denial can have and has had a powerful positive effect on cancer treatment. Most people recoil when they are given a diagnosis of cancer, knowing that it can be a death sentence. However, even with cancers for which the prognosis is grim and survival beyond a few years is highly unlikely, the vast majority of patients then respond with a strong will to live. Several cancers are treatable and even curable these days. One major reason for this progress is that both oncologists and their patients in the past had been willing to deny the reality of the situation and get actively involved in experimental treatment protocols with only a slim chance of success. Thus, while only a minority of early clinical trials have given rise to the successful treatments of today, none would have been possible if it were not for the willingness of so many patients and their doctors to deny reality and try everything possible, right to the bitter end. So while their reality denial did no good for them, it had a positive effect on the patients who came after them.

You have probably heard the common remark about someone "fighting cancer," suggesting that a positive attitude can improve the outcome. The cynical biologist who understands the molecular basis of cancer may scoff at what seem to be folktales. In fact, this is not just an observation that some oncologists make clinically but also one that has been documented. One study showed that, given the same initial diagnosis, nonpessimistic cancer patients survive a few months longer than pessimistic patients.[1] In another controlled study, breast cancer patients were assigned either to a special program of counseling and emotional support or just left in conventional care.[2] In fact, those who received the support lived an average of eighteen months longer than those who did not. So here are apparent examples of the power of optimistically denying the reality of cancer. But other studies since then have not replicated such positive results in all patients.[3] And the cynic would argue that any positive outcomes are simply because optimistic and positively thinking patients took much better care of themselves, thereby making it more difficult for the cancer to kill them. But it does not matter whether there are actually some biological mechanisms behind the impact of optimism on patient outcomes. The bottom line is that some people with cancer lived longer.

A 2012 study reported that the great majority of patients with lung and colorectal cancer have unrealistically optimistic expectations regarding the value of chemotherapy they receive.[4] The authors expressed concern that this could compromise a patient's ability to make informed treatment decisions. However, the accompanying commentary[5] pointed out that self-deception is a valuable personal coping tool, and that optimism helps us deal with the inevitability of death. So rather than focusing only on the negative connotations of these apparently misguided views of cancer patients, we should also consider the benefits. Without a positive attitude toward their situation, such cancer patients might suffer greatly from hopelessness and depression. Of course, as with all forms of optimism, one can go too far—and some patients with cancer continue to fight the odds even when they are truly hopeless, resulting in unnecessary suffering for themselves and financial losses to their families. The challenge is to find the right balance between optimism (reality denial) and objective acknowledgment of reality—without becoming overly pessimistic or depressed.

We now even have a brain-based explanation for such

irrational optimism. Tali Sharot has shown that distinct regions of the prefrontal cortex of the brain track "estimation errors," and that highly optimistic individuals exhibit "reduced tracking of errors that call for a negative update."[6] She concludes that "optimism is tied to a selective update failure and diminished neural coding of undesirable information regarding the future." And it is striking that the pathways in question (involving both the prefrontal cortex and the amygdala) are the very same ones involved in controlling fear and stress responses. In other words, the frontal lobes of your brain are much better at informing your stress pathways about good news than they are at informing you about bad news. This may well be one of the major neural mechanisms of reality denial.

Bravery and courage are considered to be positive human attributes. While bravery can verge on foolhardiness, it can still end in a success story. The history books are full of such examples. How many times have humans (individually or in groups) set out to do what seemed almost impossible and have yet succeeded? If it were not for the ability to deny reality, they would not even have tried! Denial also explains the human ability to take on dangerous and important challenges, such as exploring new territories or trying out new foods of unknown safety for the first time. So denial can also act as a buffer against our natural aversion of risk, enabling us to explore new ideas and behaviors that may eventually benefit the human population. Again, the ideal balance is to maintain this positive value of denial and yet have the ability to recognize and avoid its negative consequences.

Extreme optimism is actually a form of extreme denial of reality. My grandfather was a friend of Mahatma Gandhi, and I therefore gained a good understanding of what this great man achieved. But at the time when Gandhi first proposed it, his concept of nonviolence struck many people as foolhardy and highly unlikely to succeed. In fact, Gandhi first tried out his nonviolent approaches in South Africa and almost met his ultimate fate at the hands of the unappreciative Boers. However, the fledgling Indian independence movement realized that Gandhi might be able to use this approach against the British (who did, after all, claim to have a concept of "fair play"). He was thus recruited over to India and went on to establish a program of nonviolence there. On the face of it, this seemed to be a highly unrealistic approach given the might of the British Empire and Gandhi's lack of weapons—other than this quirky concept of nonviolent protest. As we all know, though, Gandhi finally succeeded, and while violent approaches were tried by other Indian freedom fighters it was nonviolence that finally won the day. Years later, other heroes, including Martin Luther King and Nelson Mandela (the latter was a reformed terrorist), used this technique successfully (of the three individuals mentioned, only Mandela lived to see the final success of his approach). The fall of the Soviet Union and the Berlin Wall were also at least partly related to nonviolent public protests. And now, of course, we have the Arab Spring of 2011, when ordinary people came out in a nonviolent way to protest and demand fairness and justice. As of this writing, the Arab Spring has not yet delivered on its promise and may have actually made things worse—but remember that the movements Gandhi, King, and Mandela initiated also did not seem too promising at the beginning. Anyway, none of this would have been possible without the extreme optimism (the extreme denial of reality) that allowed Gandhi to take his novel, nonviolent approach. Of course we will never know how many others tried similar approaches and failed.

Denying reality can also be a source of pleasure. When we go to see a movie or a play and get really immersed in it, we are essentially suspending disbelief and denying reality. We are also suppressing reality when we use mind-altering drugs. Some other human activities, such as competitive athletics, require intense concentration, and the key to achieving this is often to shut out everything else. Thus a sprinter on the blocks at the Olympics will say that she or he is in "the zone" and cannot hear much except the very distant roar of the crowd. Again, this is a form of reality denial. In many fields of sport, the talents of the competitors are so evenly matched that personal psychology plays a key role in success. Well-known tactics include intimidating, distracting, and outwitting opponents. But the most reliable approaches are actually variations on reality denial. Great athletes will actually visualize their success even before they compete, and psych themselves up by convincing themselves of their own invincibility and greatness. Such athletes are denying the reality of the competition they actually face while also invoking a self-fulfilling prophecy. A classic example of combining all these approaches is that of Emil Zátopek, a famous long-distance runner who won three gold medals at the 1952 Summer Olympics in Helsinki. Zátopek first won gold in the 5,000-meter and 10,000-meter runs, the races he had actually trained for. But then he decided at the eleventh hour to compete in the marathon (a much longer race) *for the very first time in his life*. His strategy was to run right alongside the British world record holder, Jim Peters, and to set a blistering pace more appropriate for the shorter races he normally ran. Once he knew that he had overtaxed Peters, Zátopek pointed out to the Englishman that he had never run this race before and casually asked whether they were running fast enough.

Shortly thereafter Peters dropped out—and all other runners behind him were psyched out as well, so Zátopek went on to win in an Olympic record time.

On an even higher plane, meditation is a human practice that comes in many varieties. While a few forms of meditation focus on gaining an increased awareness of reality, the common goal of most meditation methods is to deny reality by deliberately eliminating extraneous thoughts and focusing on only a few (in the extreme form, to focus on one single word or sound—or, even better, on nothing at all). In the final analysis, such deep meditation amounts to a deliberate and sustained effort at maximal denial of the reality surrounding an individual, both physical and mental. Perhaps this is why many forms of meditation (or "mindfulness") are shown to make humans happier and healthier. A recent journal article concluded that "mindfulness brings about various positive psychological effects, including increased subjective well-being, reduced psychological symptoms and emotional reactivity, and improved behavioral regulation."[7] Some forms of meditation, such as Buddhist zazen, go back to the time of Gautama Buddha, who taught that the way to enlightenment and avoidance of suffering was to understand that what we think of as reality is actually an "illusion of permanence." The underlying concept in religions of Indian origin is that of Māyā, which suggests that we do not actually experience the environment itself but rather a projection of it, generated by us.[8] In other words, what we think is real is actually an illusion. These ways of thinking lead to intense meditative attempts to eliminate that illusion of reality, often resulting in a greatly improved peace of mind. But we now know that reality as revealed by science is in fact the real truth. Thus all these meditative attempts to shut out the world are really deliberate forms of reality denial that make us feel better. From this point of view, the ultimate enlightenment, or nirvana, is nothing more than *complete* denial of reality. Negative as that may sound, many people believe that meditation is the highest mental state that humans can achieve. In light of our current theory, the reason seems evident. Reality (which is *not* an illusion) is often unpleasant and includes a lot of suffering and other types of negative emotions—and deep meditation takes one away from reality and calms the mind. And the reason this is possible is that we humans have a built-in ability to deny reality, an ability that we can harness and magnify.

Of course humans do reach an age at which it becomes obvious that things are going to end sooner or later. At this point the value of reality denial clicks in again, giving us the ability to accept this mortality without fear or angst. While perhaps of limited evolutionary value, this denial has practical consequences because it allows one to plan for mortality in a rational and nonemotional way. So here again is another positive aspect of reality denial. In this regard it would behoove people of a certain age to realize that although we may have amassed wealth and succeeded in career goals, we may have done some things that we regret and want to undo before death occurs. Recognizing one's mortality in this circumstance can be very beneficial, as it can be done without emotion.

We mentioned Steve Jobs earlier, and speculated that his denial of the reality of his cancer may have cost him his life. But once reality hit home, Jobs said at his Stanford commencement speech in 2005, "Remembering that I'll be dead soon is the most important tool I've ever encountered to help me make the big choices in life. Because almost everything—all external expectations, all pride, all fear of embarrassment or failure—these things just fall away in the face of death, leaving only what is truly important. Remembering that you are going to die is the best way I know to avoid the trap of thinking you have something to lose. You are already naked. There is no reason not to follow your heart." In fact, one of Jobs's most creative periods followed his realization that he might be doomed to die of cancer. On a related note, Jobs was also famous for his "reality distortion field,"[9] a term apparently coined by an Apple engineer to describe Jobs's ability to distort the proportion and scale of difficulties, making others believe that any task was possible. Apparently when Apple managers and programmers told Jobs that something was unrealistic, he would simply instruct them to "distort reality." This approach made the seemingly impossible possible after all. The resulting sense of optimism is said to have contributed both to some of the achievements of the company and to the loyalty of Jobs's colleagues.

Such realism about mortality may be gaining in popularity. A 2007 movie called *The Bucket List* depicted the adventures of two dying men who happen to meet in the hospital and decide that they're going to do certain selected things before they "kick the bucket" (there are competing theories about how this odd phrase got its meaning—a common one holds that it references a method of hanging in which a noose is tied around a person's neck while the individual stands on an overturned bucket; the bucket is then kicked away to hang the victim). This charming bucket-list concept has taken on a life of its own, and is now featured on many websites that outline different versions of the basic idea. There are even poignant sites at which individuals who know they're doomed to die from terminal cancer report courageously on their plans for a limited bucket list. Here we have humans successfully acknowledging their mortality even while not letting it get their spirits down (denying the negative emotions that arise from actual reality). In fact, they have converted knowledge of mortality into a positive, listing things that they wish to complete before they die—all the while still operationally denying that they could die the very next day. Others choose to contribute their talents to humanity right until their dying day. For example, Christopher Hitchens's final masterpiece, *Mortality*, describes his own personal experience of dying from cancer, in impassioned prose.[10] The newly emerging "Death Cafe" concept also takes the right approach.[11]

Another potential benefit of better understanding our mortality and accepting its reality could be the way in which extremely wealthy people spend their fortunes. As long as one is denying reality and mortality and feeling that life will go on forever, one would want to continue to accumulate money for the future. But the simple reality is that there is only so much one can spend in one's lifetime, no matter how hard one might try. Thus it makes sense for rich people of advancing age to first consider what they want to leave to their families, friends, and acquaintances and then decide what to do with the rest of their money—much of which they will never spend before they die. As Bill Gates and Warren Buffett have realized, it is a good idea to distribute one's wealth for the benefit of many others before one passes on. It is too bad that many other rich individuals are instead in continuing denial of the reality of their mortality, behaving as if there will be no end to their acquisition of wealth. But as the old saying goes, "You can't take it with you." And for those of a religious bent, remember that "It is easier for a camel to go through the eye of a needle, than for a rich man to enter into the kingdom of God" (Matthew 19:24).

In this regard it is worth recalling Leo Tolstoy's 1886 classic short story "How Much Land Does a Man Need?" In this tale a greedy man named Pahóm comes across a tribe called the Bashkirs, who promise him that he can have as much land as he can circumnavigate and mark from sunrise to sunset, for a relatively small sum of money. However, he would stand to lose everything if he did not return to the starting point by sunset. Pahóm was quite confident he would be a big winner in this easy task. But once he started out, Pahóm was so avaricious that he kept trying to encompass more and more land within his walking circuit. Realizing late that he might never make it back in time, he broke into a dead run and arrived just at sunset, only to drop dead of exhaustion. Tolstoy concludes, "The Bashkirs clicked their tongues to show their pity. His servant picked up the spade and dug a grave long enough for Pahóm to lie in, and buried him in it. *Six feet from his head to his heels was all he needed.*"

Perhaps once we acknowledge the reality that we each have only a very limited span of time on this planet, we may all choose to make a better go of it during the time we are here. There are many more examples of such "new" realism, in which we can continue to use our denial of reality and mortality and yet have sufficient insight into it to take proactive steps in the appropriate arenas. In fact, even for those who are not terminally ill, the realization that life is short and has a finite end can be converted into a very positive thing. It is a well-worn cliché, but very true—live each day is if it were your last, and relish every moment. As Stephen Grellet (1773–1855) so eloquently put it: "I expect to pass through this world but once. Any good thing, therefore, that I can do, or any kindness I can show to any fellow human being, let me do it now. Let me not defer or neglect it, for I shall not pass this way again."

CHAPTER 12

Explaining the Mysterious Origin of Us

I think that science has gone too far in . . . the idea that [man] is merely . . . a hastily made-over ape.

—Sir John Eccles, in *The Self and Its Brain*

But of the tree of the knowledge of good and evil, thou shalt not eat of it.

—Genesis 2:17

I f I were to coin one short phrase to encompass the concept presented in this book, it is "Mind over Reality" (the evolutionary adaptation in response to gaining theory of mind by simultaneously evolving denial of reality). Let's now consider a somewhat more academic but equally fascinating question—whether this theory can explain a long-standing mystery regarding the unusual origin of our own species. As previously explained, we "behaviorally modern humans" emerged from somewhere in Africa about one hundred thousand years ago or so, and then spread across the planet, replacing many other existing humanlike species. And from the evidence provided by fossil skeletons known to date, we seem to be a subset of so-called anatomically modern humans who were already in Africa for quite a while before that (for details see some recent books such as those by Bernard Wood,[1] Robert Boyd and Joan Silk,[2] Chris Stringer,[3] Ian Tattersall,[4] and Camilo Cela-Conde and Francisco Ayala[5]). These are really exciting but also difficult times to be discussing this topic. Even as this is being written, much intriguing new DNA, fossil, and archaeological evidence is emerging that bears on the subject,[6] including a high-quality genome from a Denisovan individual from Pääbo's group.[7] Making matters even more difficult, Richard Durbin recently opined that all the molecular clock–based timing discussed earlier may need to be recalibrated to earlier dates, based on recent data.[8]

Before the availability of modern DNA and archaeological evidence, the theory to explain the emergence of behaviorally modern humans was that of a cognitive revolution, or a "great leap forward," starting about fifty thousand years ago. European anthropologists were impressed with the dramatic findings of representational art, symbolic artifacts, and sophisticated tools in caves and other sites occupied by Cro-Magnon man in Europe. Despite emerging evidence that these humans were basically just emigrants from Africa, a theory was espoused that some dramatic genetic change altered the cognitive behavior of humans after they migrated out of Africa into Europe. Implicit in the theory was the notion that those left behind in Africa were somehow cognitively inferior. This Eurocentric view with presumably unintended racist overtones generated an eventual backlash from other anthropologists and archaeologists, who pointed to the increasing and substantial evidence for symbolic activities of behaviorally modern humans in Africa that goes back much further in time. A reanalysis by Sally McBrearty and Alison Brooks[9] cogently argued that the prominence of European findings was a consequence of the fact that Europe was the region where most of the initial archaeological work was done, and was perpetuated by the better preservation of specimens under nontropical conditions and the fact that human activities in Africa took place over a much longer time period and were spread over a much larger geographic area, making them more difficult to study than the activities in Europe. With recent discoveries of ever-more-convincing archaeological artifacts attributable to behaviorally modern humans in Africa at much earlier time periods,[10] this newer view can be summarized in McBrearty and Brooks's phrase "The Revolution That Wasn't"—a story of gradual changes beginning around two to three hundred thousand years ago in Africa that resulted in the emergence of behaviorally modern humans who later migrated into Europe and the rest of the world.

So the default theory at the moment is that the transition from anatomically modern humans to behaviorally modern humans was a gradual one, and that there was no discrete event that generated modern humans.[11] While this view of human origins is parsimonious, it is possible that we are throwing the proverbial baby out with the bathwater. Indeed, other thoughtful experts have pointed out that the proposed "revolution" could instead have occurred in Africa, at an earlier time.[12] If anatomically and behaviorally modern humans are just the same people, why is it that the former, who were extant about one hundred thousand years earlier, did not leave behind many archaeological artifacts that would indicate advanced mental behavior similar to ours, such as bead necklaces and burials with symbolic funerary items? Chris Stringer points out that while fossils such as "Omo 1 at about 195,000 years and the Herto crania at about 160,000 years seem to establish an early modern presence in Ethiopia, and the Guomde fossils from Kenya might even take that presence back to 250,000 years ago," there is "no evidence of symbolic burials older than the early modern examples known from Skhul and Qafzeh at about 100,000 years."[13] Meanwhile, Ian Tattersall emphasizes that "the earliest anatomical *Homo sapiens* appear right now to have been cognitively indistinguishable from the Neandertals and other contemporaries"; and that "the archaeological record seems to indicate a very late and essentially unheralded arrival of symbolic consciousness in just one lineage of large-brained hominid."[14] In the same vein, Colin Renfrew opines that "genetic change clearly played a significant role in the formation of our species. Perhaps it continued to do so as late as about 75,000 years ago, when we see the emergence of the more modern behaviors associated with the human revolution."[15] No one currently has data to conclusively resolve this puzzle.

One subset of theories suggests that there was a population "bottleneck," in which environmental factors caused a constriction of anatomically modern human populations into a small group that eventually emerged as behaviorally modern humans—while everyone else died out, simply by chance. Supporting this notion is DNA data suggesting that all humans on the planet today came from an effective population size of around five to ten thousand breeding individuals. And there are even candidate mechanisms for the presumed bottleneck. Besides the impact of the huge climate swings that were occurring due to the comings and goings of the great ice ages,[16] there was Mount Toba, a "supervolcano" that erupted in Indonesia around seventy-four thousand years ago. This cataclysmic event (more than one thousand times the intensity of some of the eruptions you may know about, such as

Mount Saint Helens) dispersed ash as far as Greenland, Japan, and Australia, laid waste to much of Southeast Asia and South India, and likely had a devastating impact on climate throughout the world.[17] The anatomically modern humans in Africa would have no doubt felt the impact of this eruption. Stanley Ambrose suggests that their populations may have shrunk dramatically as a result of this eruption, which perhaps generated the bottleneck and also then assisted in the dispersal of the human population.[18] Another (not mutually exclusive) possibility comes from my own work, which examines evidence that the human immune system may have had a catastrophic interaction around one hundred thousand years ago with certain bacteria that could kill fetuses and infants.[19] Although we can see the "molecular ash" from this proposed event in our genomes and immune systems today, the timing is less definitive than that suggested by volcanic ash. Regardless of whether these and other factors might explain the presumed bottleneck, we also have to acknowledge alternative models, which suggest that the apparent bottleneck was just a random feature of a widely dispersed and small population.

How about the mystery of what happened once we left Africa? The original assumption had been that we behaviorally modern humans not only had mental abilities superior to those of the others already present outside Africa but also that we were a genetically different, distinct, "reproductively isolated" species unable to breed with them. But, as we mentioned earlier, DNA data show that we did in fact interbreed with other hominins, such as Neandertals and Denisovans, when we encountered them outside Africa. And recent genomic evidence from Michael Hammer and Sarah Tishkoff indicates that humans who stayed behind in Africa also mated with some other coexisting humanlike species.[20] While such matings transferred a small quantity of DNA into many of our bloodlines, the intermixing does not seem to have had any major impact on our biology other than giving us a few important genes that likely protected us from local infectious diseases.[21] In other words, there is as yet no evidence for "positive selection" having worked on most of this nonhuman DNA (certain genes involved in muscle function and protection from ultraviolet light do show evidence for selection in the Khoe-San populations of Africa, who appear to share the deepest roots with the rest of humanity).[22]

Despite this, a common statement we hear in the media is that "Two to four percent of genes in Europeans and Asians came from Neandertals" or that "Four to eight percent of genes in Melanesians came from Denisovans." Given that there are more than twenty thousand genes in the human genome, this should amount to about several hundred genes that we incorporated from these extinct evolutionary cousins. But there are so far only three examples of actual Neandertal or Denisovan versions (alleles) of genes that have clearly been incorporated ("introgressed") into the genomes of present-day humans,[23] and these are the ones we already mentioned as affecting the immune system. It is still too early to tell what the final count of introgressed genes will be. For the moment the data confirms that we were just a subspecies of humans that could interbreed with other humanlike lineages that were around at the same time, and that we were capable of acquiring useful genes from them in the process. So one wonders why we do not see more examples of incorporation of Neandertal and Denisovan versions of genes into modern human genomes. It could have been just chance. On the other hand, it is plausible that there was a psychological or cognitive type of barrier or difference that greatly limited the possibility, unless the genes were really crucial for survival, as were the few we acquired in our immune system.

To put the question a different way: If behaviorally modern humans coexisted with so many other cross-fertile subspecies for tens of thousands of years, then why did we not hybridize (mix up our genes and genomes) with them much more? Why are present-day Europeans not 30 percent Neandertal? Why are present-day Southeast Asians not 30 percent Denisovan? And why are present-day Africans not 50 percent different from everyone else who left Africa? More data is needed, and we also cannot rule out relative infertility of the hybrids. But for the moment I suggest it is the "b" in *behaviorally* modern humans that explains things best. My assumption is that we were a biologically cross-fertile but *cognitively distinct* subspecies of anatomically modern human. And if "Mind over Reality" was the key breakthrough of behaviorally modern humans in Africa, it was likely mediated by a complex series of genetic events. So any first-generation progeny of our matings with Neandertals or Denisovans would be largely missing this ability and such progeny may not be well assimilated into the human group into which they might be born.

To recap, behaviorally modern humans are nothing more than a subspecies of *Homo sapiens,* which included some other branches of anatomically modern humans in Africa, Neandertals in Europe and Denisovans (probably in the East), as well as other subspecies that probably coexisted and have yet to be discovered. If this is true, then why did we not fully interbreed with one or more of these other subspecies despite an opportunity to that spanned *tens of thousands of years* (a period many times longer than all of recorded human

history)? Conversely, since we did have contact with Neandertals and presumably interacted with them culturally, why
did they not adopt all our practices if they were already mentally capable of doing so? Perhaps it was just chance that
kept all these subspecies apart. But then why stay apart for so
very long? Perhaps it was sexual selection and mate choice,
as happens with other subspecies that can coexist in the same
environment. Given the human propensity (especially among
males) to raid mates from adjacent tribes, this seems unlikely.
It seems more likely that the progeny of such mating was either of much-reduced fertility or did not have the behavioral
and cognitive abilities that would include them in the future
gene pool of reproducing humans. The following discussion
assumes the second possibility.

A scenario that could represent a suitable intellectual compromise is as follows. In this model anatomically modern
humans were already present in Africa and areas near it for
at least one hundred thousand years and had earlier shared
a common ancestor with the precursors of Neandertals who
evolved in Europe and the Denisovans who had possibly migrated farther east. Although intelligent and likely capable of
significant intellectual prowess, these subspecies of humans
were all held back from further cognitive advances because
they did not have full ToM—and each time an individual
evolved the genetic ability to gain this potentially valuable
asset, she or he faced the intolerable and maladaptive psychological barrier of understanding mortality. We behaviorally modern humans may then be a subset of anatomically
modern humans who finally managed to break through this
barrier by simultaneously acquiring reality denial, a process
likely involving multiple genetic changes over several generations. In this scenario, individuals who achieved this transi-

tion at the same time would be much more likely to mate with each other for sociocultural reasons and because they were more capable of relating to each other mentally. This process could relatively quickly result in the *cognitive and cultural isolation* of a group of humans who constituted our ancestors. As we then spread across Africa or emigrated from Africa to other parts of the world, we would have encountered other anatomically modern humans as well as other related species, such as Neandertals and Denisovans. Individuals would be able to mate between these subspecies and have progeny. However, the special combination of full ToM and reality denial must have required that many genetic changes be inherited at the same time in the same person. Thus the chance that any progeny of a hybrid mating would simultaneously and fully inherit both abilities was very low. In other words, the hybrids were likely "cognitively sterile," in that they lacked full ToM and/or reality denial. If so, such progeny would not likely be accepted and assimilated into the primary human group. Over the next tens of thousands of years, we would thus see a gradual replacement of all the other Homo sapiens subspecies by us, the cognitively superior behaviorally modern humans.

In keeping with this, the picture emerging is that behaviorally modern humans were indeed the ones routinely capable of mental activities that would require full ToM, such as ceremonial burials, crafting personal ornaments, and making representational art. In this compromise model there was neither a great leap forward nor the complete absence of a revolution. The "soft revolution" that occurred was psychologically based, but also had sufficient genetic underpinnings for a distinct cognitive phenotype (and corresponding neural endophenotype), minimizing the possibility of broad homogenization with other human subspecies that coexisted. Incidentally, full ToM is also needed for attaining the "knowledge of good and evil," which the Bible says is the reason Adam and Eve were driven out of Eden.

While I cannot conclusively prove it, the Mind over Reality transition can thus account for the mysterious late emergence of behaviorally modern humans from anatomically modern humans; it can also account for the fact that we effectively replaced everyone else. Indeed, if you look back in chapter 4 at the examples of human abilities that require ToM, you can see a general match with the archaeological artifacts of behaviorally modern humans and a mismatch with those of their precursors. But what of the remarkable prowess of these other subspecies with regard to stone tools, ochre pigment, and the like? Perhaps among those populations were some people who were like Dustin Hoffman's autistic savant character in the movie *Rain Man*. Such individuals would have been quite intelligent but still incapable of fully understanding the minds of others. Thus, despite their intellectual prowess, the production and cultural propagation of novel technologies and ideas would have been limited by their lack of ability to teach, be taught, or pass along information to third parties. Note that some autistic individuals today who are without established ToM are still capable of producing some remarkable representational art. So symbolic art per se is not an indication of ToM. The more persuasive evidence for ToM consists of burials accompanied by funerary goods (which implies an understanding of death) and bead necklaces (which implies an understanding of the minds of others who can appreciate the ornaments).

Perhaps what was holding back even these very intelligent individuals would have been the lack of a full ToM—the inability to fully understand the minds of others around them. I suggest that the critical transition then took place in our immediate ancestors, allowing them to hold and sustain a full ToM and to propagate these abilities in subsequent generations. These abilities would have obviously given this group some great advantages. This would also explain why there was no extensive assimilation of others from coexisting subspecies. Besides the difficulty in transmission of what is presumably a very complex genetic trait, any progeny resulting from Neandertal-human mating who ended up in a behaviorally modern human group might not have been successful in further mating because of the relative or absolute lack of ToM and the inability to interact well with others. So the scattered DNA fragments from these other subspecies found in some of our genomes today may be simply a form of "contamination" left over from those failed attempts to assimilate. And what happened to the hybrids? Maybe they also failed to be assimilated back into Neandertal and Denisovan societies. Maybe they account for some of the contemporaneous fossils that Erik Trinkaus and others have described that show some combinations of modern human and Neandertal features.[24] Perhaps this is why some of the nonhuman artifacts of that era of potential hybridization (such as Châtelperronian[25] and Bohunician[26] ornaments) show some features similar to those of contemporaneous Cro-Magnons.[27]

We may never know for sure unless new data comes in to resolve various issues. But although it is not possible to directly test this model right now, all knowledge to date is consistent with it. If correct, we would predict that modern humans have evolved specific neural mechanisms that allow us to deny reality despite understanding it. Candidates for such pathways would involve fear responses and fear conditioning—such as those that extend from the "executive" regions of the frontal lobe cortex to the amygdala, a small, deeper-brain structure associated with fear and stress responses in many studies.[28] Interestingly, these are the very pathways involved in the optimism bias, which we have suggested is simply a variation of reality denial. And if and when it becomes possible to test for this bias in nonhuman animals, I predict that it will be missing. In this regard it is interesting to note that the amygdalae of great apes (who lack full ToM),[29] of people with autism spectrum disorders (who have aberrant ToM),[30] and even of individuals who have Williams syndrome (who seem to have excessive sociality and lack of fear of strangers)[31] show differences from those of typical humans.

Another prediction is that if and when DNA samples are successfully isolated from African anatomically modern humans, they will fall into two categories. The first group will be individuals whose genomes have considerable differences from ours, and the second will be those that appear to be our ancestors. And the latter fossils will be the only ones associated with the archaeological signature of activities that imply the presence of a full ToM. Last but not least, this theory can also help explain Alfred Russel Wallace's conundrum. How is it that individuals who originated from Africa one hundred thousand years ago already had the ability to perform modern human mental functions and behaviors even though they had no use at the time for the vast bulk of these functions and behaviors? Perhaps Wallace was right after all, and the modern human mind indeed evolved as "an instrument...developed in advance of the needs of its predecessor" in a one-of-a-kind Mind over Reality transition.

CHAPTER 13

Future Directions

> Long before having arrived at this part of my work, a
> crowd of difficulties will have occurred to the reader.
> Some of them are so grave that to this day I can never re-
> flect on them without being staggered; but, to the best
> of my judgment, the greater number are only appar-
> ent, and those that are real are not, I think, fatal to my
> theory.
> —Charles Darwin, in *On the Origin of Species*

Danny and I have presented a theory supported by multiple lines of available evidence and not negated by any
known facts. I have expanded on several aspects of the theory and added some implications for the recent origins of our
own species. In this closing chapter I consider future directions for exploring the theory in the context of some of its
current limitations. This will ensure that any dialogue resulting from this book is fully informed regarding the issues at
hand. But first let me reassure the reader that this book is not
meant to present an "umbrella hypothesis"—a hypothesis
that attributes multiple phenomena to a single cause.[1]

Almost every one of us has a theory about human origins,

be it scientific, religious, philosophical, or simply involving folk logic. This seems to be part and parcel of being human. There are also many formal academic theories concerning human origins, such as "man the hunter," "man the tool-maker," the "cooking hypothesis," the "cooperative breeding hypothesis," the "social brain hypothesis," the "Machiavellian intelligence hypothesis," the "cultural intelligence hypothesis," the much maligned "aquatic ape theory," and so on. In each instance, scientists consider one plausible mechanism that might have contributed to human origins and test it rigorously against what is known about uniquely human features. Many of these theories greatly assist our understanding of human uniqueness, and, taken as a whole, they are extremely helpful. However, there is a tendency for some other scientists to then go further and conceive of an umbrella hypothesis, in which one of these interesting theories is applied to *all* aspects of human uniqueness and through which scientists search for *all* potential connections. As this process continues, there is a danger of getting into "just so" stories, which are nothing more than pure speculations that happen to sound good. The theory we have put forward might appear to be doing the same. After all, many aspects of human uniqueness are referred to here in the context of this theory, and plausible connections are suggested. However, this theory is *not* meant to be an umbrella hypothesis. We are primarily talking about a *single critical step* in human evolution, in which an unlikely confluence of mental and psychological events and mechanisms allowed us to transition from simple self-awareness and rudimentary ToM to full ToM. And, as suggested earlier, this theory of a psychological evolutionary barrier can be placed in the context of several other theories to help us in understanding a key late phase of human evolution, the emergence of behaviorally modern humans.

We must also stop and ask if our basic hypothesis is falsifiable. In the early days of science most scientists were naturalists, who simply observed natural phenomena, collected information, and drew conclusions that seemed reasonable from their findings—without elaborating a hypothesis in advance for testing. The quintessential examples of successful naturalists were Darwin and Wallace, who derived the theory of evolution via descent by natural selection mainly by collecting a large amount of information from many sources and diligently analyzing it. Of course naturalists can (as Darwin and Wallace did) test their theories against related facts discovered by other naturalists to see if they still hold up. But such early naturalists often could not prove much with certainty. As the methods of science progressed, it became feasible to conduct experiments to confirm or refute ideas originating with naturalists, and the best experiment was one in which a clear hypothesis was available to be tested precisely. And unless a hypothesis is falsifiable (in other words, possible to prove wrong in an experiment), it is less useful. This logical progression of the scientific method reached its zenith with the philosopher Karl Popper,[2] who suggested that one should not even bother to make a hypothesis unless it is falsifiable. This thinking now dominates many scientific fields and makes it difficult to justify studies that do not have an experimentally testable or fully falsifiable hypothesis. If Charles Darwin were to write a research grant application today, it would be highly unlikely to be funded! Yet history tells us that many scientific revolutions (so-called paradigm shifts) do not necessarily arise from the study of preexisting testable hypotheses.[3]

Regardless, is our Mind over Reality hypothesis falsifiable? One cannot think of an ethical way to make it so with the tools and knowledge available today. This may change in the future, with further studies of the psychology, genetics, and neurobiology of humans and those of our closest living evolutionary relatives (and with better studies of the archaeology of human ancestors). For the moment all we can say is that the theory *fits all available evidence* and that there is *no strong evidence against it*. Indeed, can any other theory explain why, despite the existence of some form of self-awareness in warm-blooded social mammals and birds with high intelligence, only one very recent species over the course of tens of millions of years seems to have finally attained a full ToM? A reasonable explanation is that there has been a barrier preventing all these other species from progressing further. Awareness of reality and mortality could have been such a barrier, and at the moment we can think of no others. So while we do not have a falsifiable hypothesis, we began with a plausible one, and I believe that we have elevated it to the status of a justifiable hypothesis. Also, there are several testable predictions arising from our hypothesis. To those who still remain skeptical, remember that absence of definitive evidence is not definitive evidence of absence. Do also remember that when something happens *only once*, it could have taken place via the most improbable series of events rather than the most logical path. An analogy can be made with a murder mystery, in which no amount of logic can substitute for uncovering clues that explain a singular event.

As explicated in the main body of this book, the Mind over Reality hypothesis claims that achieving and sustaining a full theory of mind during evolution required simultaneous attainment of reality denial—a highly unlikely combination that happened only once during biological evolution on this planet, at a point close to or simultaneous with the emergence of behaviorally modern humans. It is likely that some experts would tend to disagree a priori when presented with a novel theory that goes against the conventional wisdom—the orthodox view, which assumes that human mental abilities gradually evolved step-by-step via typical natural selection, assisted perhaps by gene-culture coevolution. It is certainly true that some of our supporting data are less than definitive, that many specific questions remain unanswered, and that there are many issues arising. To avoid distracting the reader with such issues throughout the book, I have chosen to present them all in one place in this final chapter, and to also suggest the additional ideas and future directions that emerge. The list is doubtless incomplete.

1. The mirror self-recognition test may not conclusively demonstrate true self-awareness. This is certainly a reasonable argument, one that has been advanced by some purists. Further studies are needed to determine if the test is simply demonstrating an advanced version of body image awareness (which many animals and birds must have) or if it represents true recognition of oneself as an individual mind (or recognition of one's own personhood). However, both the body self-examination behaviors that occur following successful administration of the mirror test and the chimpanzee responses to circus mirrors described by Derek Denton suggest that the test is indeed a valid proxy for self-awareness. Regardless, alternative methods to detect this cognitive state are clearly worth pursuing to see if the very same species test positive.

2. The experimental evidence cited for self-awareness in some nonhuman species is not definitive. This is certainly true, and experimental replications and further evidence are needed in other species, except possibly in chimpanzees. But even if future studies refute all current opinions and show that this ability is actually limited to our closest evolutionary cousins, it still does not change the primary thesis that the Mind over Reality transition occurred uniquely in humans.

3. The self-awareness phenomenon in various species may not result from the same neurocognitive processes. This is in fact very likely to be the case, given the apparently independent emergences of self-awareness in several disparate evolutionary lineages. If so, this actually lends even *more* credence to the current theory, raising the question of what barrier has been holding back all these *independently evolved, self-aware* species from moving on to the obvious next step, which is simply becoming aware of the self-awareness of others. We suggest that the psychological evolutionary barrier common to all such species consisted of negative and maladaptive reactions to awareness of mortality and reality, following each initial emergence of ToM.

4. Theory of mind is not a clearly definable concept. As we have discussed, there is indeed no shining bright line between self-awareness and theory of mind. It seems that they are really just two points on a continuum, which goes from incipient self-awareness to full self-awareness to rudimentary theory of mind to full theory of mind. The same could be said of alternative terminologies for these phenomena, such as multiple levels of intentionality, which likely also can be placed on a continuum. Regardless of such details, available data indicate that humans are unusual in *where we reside on this continuum*. Of course, not all the other candidate animals mentioned have been thoroughly studied, and it may be that some form of this humanlike ability exists in some other species.

5. *Theory of mind and reality denial in humans could have evolved piecemeal, working together little by little, over many generations.* We cannot rule out this possibility. But the intermediate stages would likely have been rather unstable, due to the risks resulting from denial of practical realities and/or the anxiety resulting from the incipient awareness of mortality—as well as possibly suicidal depression and/or pathological lying. And the fact remains that this unusual combination has been very difficult to find in other animals.

6. *A form of reality denial is present in other animals.* It is true that other intelligent animals can show cognitive avoidance, inattention to immediate realities, or suppression of reactions to repetitions of unpleasant but nondangerous stimuli. In fact, these mechanisms may represent the evolutionary precursors to the human reality denial discussed in the current theory. However, humans seem to be the only species with such an advanced understanding of reality and mortality that we need ongoing denial in order to tolerate that understanding.

7. A rational human being can deal with mortality fears simply by looking at facts and statistics regarding death risk. This is indeed the case with present-day humans. However, we are the *final* products in our proposed evolutionary transition. The ability to understand mortality had to first occur in a few individuals in an isolated situation. And given the conditions that humans were under at the time, evidence of mortality risk (e.g., the deaths of mothers in labor, infant mortality, accidents, starvation, attacks by predators, and so on) was likely to always be present. And the first individual making the transition would have been alone in his or her fear, with no one to discuss it with, and would not yet have been capable of internally rationalizing it. Given the severe anxiety, fear, and depression this would generate, such an individual would likely fail in the evolutionary competition to pass on his or her genes.

8. The biological mating urge is so strong that knowledge of mortality would have not been sufficient to suppress it. This is indeed possible. But our theory is not suggesting a complete loss of reproductive fitness when faced with understanding of reality and mortality. Rather, it posits a significant *decrease* in fitness such that the *probability* of this rare combination of psychological abilities being sustained and becoming genetically fixed in a cohort of individuals in a given species was extremely low. Remember that multiple genes must have been involved, and that these genes would have been engaged in a complex interaction with the environment.

9. *Neandertals may have had a full theory of mind.* We have presented some of the evidence against this. However, it is true that some perceived Neandertal behaviors (such as caring for the injured) make one wonder how close they were to being mentally like humans. But, as we have discussed, the capacity for "mind-reading" thoughts may be distinct from that for reading emotions (which would be more important for empathy). And even after meeting modern humans and having the opportunity to interact with us for more than ten thousand years, Neandertals did not go on to mate freely with us, nor did they show strong archaeological evidence of full theory of mind, such as representational art, artifacts of trade, personal ornamentation, sophisticated grave goods, and so on.

10. *Production of ochre pigment and long-range transport of obsidian tool-making materials predate the origin of behaviorally modern humans.* This is true, and these activities suggest some degree of symbolic thought or the ability to think ahead. And the perpetuation of these cultural abilities may have required some level of instruction in order to be passed on. We suggest that individuals with savant-like capabilities were present within these older populations prior to the Mind over Reality transition, i.e., prior to the origin of behaviorally modern humans. But the fact that these artifacts remained largely unchanged for tens of thousands of years suggests a difficulty in bootstrapping these behaviors, a difficulty that resulted from a lack of full theory of mind—based interactions, which are needed for active teaching, group instruction, long-distance communication, and trade.

11. Advanced tool production prior to modern humans must have required some degree of teaching, verbal communication, or, at minimum, active demonstration. The advance planning needed to prepare some types of stone tools would indeed imply a sophisticated manner of thinking. We do not rule this out, and are not suggesting that a full theory of mind emerged very suddenly. Earlier stages of theory of mind may well have occurred over the last few hundred thousand years in various hominin lineages (including Neandertals and anatomically modern humans)—but just did not progress further until the full denial of reality could also fall into place, which it did in our lineage.

12. Autism spectrum disorders cannot be used as a direct proxy for our ancestral cognitive state prior to the emergence of modern humans. This is true. But we are simply considering these disorders as variations of nature that reveal the consequences of not having a full theory of mind. Notably, the technical and artistic achievements of some individuals with mild forms of ASD and savant-like special talents may have even facilitated the emergence of modern behaviors in the ancestral population because they were imitated by others in that population.

13. There are as yet no defined neural pathways to mediate the proposed evolutionary changes. Our theory would be strengthened if one could find unique correlated evolutionary changes in specific components of the human brain. Our knowledge of the brain is insufficient at the present time to make such direct correlations. But current findings allow one to speculate a bit. Several investigators have studied the human brain by functional imaging during simulated processes of "mentalizing" (imagining the minds of others) and noted activation of several specific regions, mostly in the frontal cortex or near it.[4] Of course, it would be overly simplistic to state that these are the "seat of the theory of mind," especially since there are no reported examples of localized strokes selectively eliminating ToM only. But one can say that this region of the brain is likely involved in attributing mental states to others. Meanwhile, studies in so-called existential neuroscience show that prefrontal cortical areas also tend to be activated during exposure to death-related statements.[5] And it is intriguing that the prefrontal cortex is involved in the optimism bias, in which this region of the brain sends updating signals to the amygdala (a deep structure in the brain that tracks fear responses).[6] But the prefrontal cortex is of course involved in many other activities, and one must be cautious in coming to any sweeping conclusions. However, we can speculate that the rare coincidence of theory of mind and denial of reality emerged because they became physically intertwined in the prefrontal cortex and amygdala during the evolutionary development of the human brain. And, as suggested earlier, the process need not necessarily have involved generation of new neural mechanisms for reality denial. Rather, the emergence of the neural substrate necessary for a full ToM could have incidentally and partially altered pathways from the prefrontal cortex to the amygdala that mediate normal fear responses and fear conditioning, allowing a blunting of such responses to reality. It is also intriguing that some differences between humans and apes have been reported in the prefrontal cortex and amygdala of both species,[7] and that the human amygdala is altered in cases of autism spectrum disorders.[8]

Motivated readers can doubtless come up with more questions and issues arising from our theory, as well as ideas for future directions of research. I would welcome hearing about them as the discussion and debate on this theory go forward.

Epilogue

The past is but the beginning of a beginning, and all
that is or has been is but the twilight of the dawn.
—H. G. Wells, in *The Discovery of the Future*

We have come to the end of this highly improbable
story. A single brief chance encounter between two
individuals from very different backgrounds but with a common intellectual interest led to this attempted synthesis of a
potentially big concept and its wide-ranging implications for
the origins and future of humanity. The cell biologist and geneticist Danny Brower, who had the initial germ of the idea
and began this project, has passed on. And I, a physician-scientist turned anthropogenist who most unexpectedly
picked up this baton, have tried my best to take it to at least a
preliminary finish line. Neither of us was or is a widely published expert on the main subjects at hand, the evolutionary
origin of humans and the cognitive psychology and evolution
of the human mind. So our relative naïveté on some issues
likely shows through in this book. It is for the same reason
that I have not attempted a detailed citation, exposition, and
interpretation of the vast and excellent peer-reviewed litera-

ture on all the relevant topics. Rather, I have relied on the numerous lectures on many aspects of anthropogeny that I have been privileged to hear at CARTA meetings—and also on interpretations in books and reviews by others, which are cited to the best of my ability, along with a few original sources. And when they are approached with a critical mind, Internet sites like Wikipedia, Google Scholar, TED.com, and so on are extremely valuable sources of information. It is safe to say that if it were not for all the remarkable resources mentioned above I would never have taken on this challenge, let alone completed it. And, of course, none of this would have happened if I had not had a chance meeting with Danny on that fateful spring day.

Experts on some issues involved may find fault with some of the specifics and details of our theory as presented. Moreover, unlike an ideal scientific hypothesis, this one is not formally subject to falsification, as there appears to be no definitive test that could prove it wrong. Thus I have relied instead on presenting "one long argument,"[1] in which I have sought to find all facts that fit with the theory. There are numerous such findings discussed throughout the book. I have also searched hard for the "ugly facts" that would "slay this beautiful hypothesis."[2] As far as I can tell there are no such facts. It is now up to others to consider the ideas put forward and follow through with attempts to either falsify or support what appears to be a justifiable hypothesis. I have also frankly discussed some of the unresolved issues and the future directions arising from them.

This book began with a specific idea initiated by Danny Brower—an attempt to explain why so many highly intelligent species have not progressed to developing an awareness of the self-awareness of others (a full ToM) despite all the obvious benefits of doing so. The suggested explanation is that this was not possible without *simultaneously* becoming able to deny the risk of mortality, a very rare combination. But while the book and the theory began with an emphasis on denial of mortality, the concept is just the tip of the iceberg of the human ability for denial of many realities, a denial that goes against any and all inconvenient facts. Many of our other human frailties and mental quirks seem to be variations on the same theme. In later chapters we therefore turned to broader considerations about the human ability for denying reality and finally expanded our discussion to the implications for our species and even for the climate of our planet. We hope that humanity will heed this wake-up call before it is too late for our biosphere and for our civilization.

If humans are so excellent at denying reality, we must of course ask what "reality" is to begin with. Michael Shermer suggests a process called *belief-dependent realism*, in which "reality exists independent of human minds, but our understanding of it depends upon the beliefs we hold at any given time."[3] Chris Frith emphasizes that all our knowledge of both the mental and physical worlds must come to us through models created by our brain.[4] In keeping with this, Stephen Hawking[5] suggests that "our brains interpret the input from our sensory organs by making a model of the world. When such a model is successful at explaining events, we tend to attribute to it, and to the elements and concepts that constitute it, the quality of reality or absolute truth." So our individual realities are actually nothing more than models that our own brains create of things we can perceive, combined with knowledge and memories we have accumulated—which we come to believe are real. In other words, each of us possesses our own unique and distinct reality, which no one else can ever really experience. Many aspects of our own reality are things we find unpleasant (such as mortality), and our ability to suppress or deny these aspects is a very important component of being human.

So at the end of the day we humans are a vast collection of individual realities, which overlap with each other while we are alive and interacting with one another, like so many ripples in a vast pond. Once I am dead my reality will be gone from this world, from my own perspective. What if anything might lie beyond death is not a matter addressed in this book. But bits and pieces of our individual realities can persist after we are dead, within the realities of others who remember us and within the things we said, did, or wrote. So you can actually be immortal by living on within realities of others, even after you are no longer alive. Consider for a moment that my own reality collided with that of Danny Brower for only a brief instant in time. And yet that single chance intersection of two realities initiated a cascade of events that eventually resulted in this book. Thus it is that certain aspects of Danny's reality will live on not only in the minds of those who cared for him, but will also enter the realities of many others who read this book and cogitate on its implications.

This Mind over Reality transition required a highly unlikely combination that happened only once on this planet, and I suggest that it approximately coincided with the emergence of behaviorally modern humans. Regardless of whether or not this theory can be formally proven in the near future, many of the corollaries that arise from it seem to have immediate relevance to the human condition. Our uncanny ability to deny reality (except when we are denying our reality denial!) is a powerful mental skill that seems to be unique to humans and critical to our emergence as the only species that is fully aware of our reality and mortality and yet able to tolerate this knowledge. As we have discussed, this ability has both positive and negative value. On the one hand, we blithely deny the reality of critical issues at the personal, familial, social, cultural, communal, national, international, and global level, at our peril (and of all these, climate destabilization denial is the most dangerous). On the other hand, we can acknowledge our denial and address these important issues. But we can still continue to deny reality enough to be optimistic and confident, allowing us to move forward with ideas and projects that may seem impossible to the true realist. Last but not least, we humans are capable of deliberately distorting, modifying, and even changing some realities for either good or evil purposes.

From a scientific perspective, our theory can contribute to an explanation of human uniqueness. No one would argue that without full ToM, many things that make us apparently unusual as humans would not exist—just consider the consequences of the social communication deficits in autism spectrum disorders. The hypothesis espoused here (a psychological evolutionary barrier that was very difficult to cross) can also help to explain Wallace's conundrum, the paradoxical situation in which the human brain appears to have been pre-evolved to do so many amazing things even though evolution does not "plan ahead." Once this barrier was crossed by simultaneous denial of reality and mortality, full ToM would have immediately given many new mental opportunities to humans, allowing us to take maximum advantage of our long period of slow brain growth after birth, which occurs even while interacting with many other individuals who also have a full ToM and are continuously transmitting information to us, driving the ongoing cultural evolution of our species.

Perhaps this is how an odd "naked ape" with largely degraded physical and sensory abilities, a dangerous childbirth process, and grossly helpless young managed to narrowly escape from being an endangered species and eventually took over the planet using its mental skills. It's as if our special passport were stamped with MIND OVER REALITY. We can now use this powerful and unique combination of mental abilities to either make things better for ourselves and the only planet we call home—or we can continue to misuse them and sow the seeds of our own destruction. As for reality denial, the time has come to fully recognize this peculiar, paradoxical, and potent quirk of the human mind and acknowledge its downsides, even while making use of it to benefit all of humanity as well as our biosphere and our planet.

Coda

Any theory that makes progress is bound to be initially
counterintuitive.

—Daniel C. Dennett, in *The Intentional Stance*

S ome readers may wish to skip this final section. But those
who wish to revisit and ponder the underlying concept
in this book may find it useful to read the following formal
summary of the Mind over Reality theory for the evolutionary emergence of the human mind.

- Each species on the planet is unusual in its own right.
  Humans are rather unusual in the way our brains
  work. One key feature is that we can put ourselves in
  the mental shoes of others, thus understanding each
  other's minds, thoughts, and actions and then easily
  imitating them if we wish to. This ability has allowed
  humans to transmit complex and increasingly sophisticated cultures and technologies to future generations,
  aided by various means of communication, most especially language.

- Many other warm-blooded, socially complex mammals and birds are also capable of remarkable feats of intelligence and cognition. Available evidence suggests that some such species are similar to humans in being self-aware—like us, they seem to be aware of themselves as individual minds and thus perhaps are able to understand their own personhoods.

- Despite such abilities, none of these species appears capable of fully understanding the self-awareness of others of their own kind—or, further, capable of understanding that others are aware of their own self-awareness. This ability to fully attribute mental states to others is sometimes called full theory of mind (ToM). Such a capacity so far appears to be uniquely human and is critical to many of the unusual features of our minds. We humans go further, in that we can hold an extended ToM of many individuals—and even, based on secondhand information, understand the minds of others we have not met.

- Without this full and extended ToM, much of what the human mind can and has achieved would not be possible. Why is it that only humans achieved this powerful ability, while other species with self-awareness did not? Broadly considered, there are two possibilities.

- The first possibility (the conventional view) is that the neural mechanisms required were most unusual, involving an extremely rare combination of molecular and cellular changes in the brain, which only happened once during evolution. In this logic, it is assumed that full

ToM was an immediately useful attribute, providing humans with a unique and powerful ability to understand the minds of others. Conventional natural selection would then take over, increasing the likelihood of such individuals passing on their genes. Additional improvements in ToM would then occur by further positive selection. This is the currently popular theory, and many candidate mechanisms have been explored and discussed. However, no such mechanisms have yet been clearly defined as being unique to humans, in comparison to our closest evolutionary cousins.

- The alternate possibility (the unconventional, contrarian view presented here) is that the initial acquisition of full ToM (becoming fully aware of the self-awareness and personhood of others) actually had immediately *negative* consequences for the individual. Thus the capacity to sustain and propagate this ability within any species may have been repeatedly blocked by a *psychological evolutionary barrier* that only humans finally transcended. Assuming the latter possibility is correct, what might this barrier have been?

- All animals have built-in reflex mechanisms for fear responses to dangerous or life-threatening situations. An individual of a given species who attains full ToM for the first time will also for the first time fully understand the personhood of other individuals of its own kind. At first glance this would seem a positive attribute, allowing one to control or even deceive others. But upon then witnessing the death of another individual of its own kind, the one with full ToM would also become aware of his or her own mortality and death risk (mortality salience).

- Given the preexisting built-in mammalian reflex mechanisms for reacting to death risk, this new knowledge of mortality should induce an extreme degree of fear. A present-day human can deal with such fears by rationalizing and calibrating the risk. But the first individuals to understand mortality would not be able to rationalize or understand the fear—because there would be no prior knowledge base to go on and also because there would not necessarily be any other individuals with whom to commiserate or discuss the death risk. A deep fear and anxiety would result, perhaps progressing to depression and even to suicidal tendencies.

- Thus, rather than take advantage of the obvious benefits of full ToM, the individual is more likely to avoid all potential death risks, including those involved in competition with others for resources and mates. In other words, individual *personal survival* would take priority over behavioral drives that typically ensure *species survival.* The individual would thus be unlikely to succeed in the competition to pass on his or her genes to the next generation—there would be initial negative and not positive selection for this trait.

- Beyond this psychological evolutionary barrier, another factor limits the possibility of establishment of full ToM in a species. There is no evidence for a single gene or molecule responsible for this ability in humans. Rather, it is likely that this complex ability results from optimal interactions among a suite of genes and molecular and cellular mechanisms. Thus, in order to pass on this ability to the next generation, it would likely be necessary for more than one individual of both sexes to develop this ability at about the same time, and the ability would eventually need to become genetically stabilized in a population of individuals.

- Taken together, the above combination of factors makes it highly unlikely that full ToM could become easily established in a species. It is quite possible that such episodes of appearance and disappearance of full ToM have been occurring in many species for tens of millions of years, until humans finally breached this psychological evolutionary barrier. How did humans achieve this?

- One plausible mechanism that would allow escape from this psychological evolutionary trap would be for individual(s) with ToM to simultaneously attain the ability to deny mortality and death risk. This would require neural mechanisms to diminish the deep fears arising from understanding mortality.

- However, simultaneously evolving a highly selective neural mechanism for suppressing fear of mortality would be difficult. A general mechanism to deny reality (including the risk of mortality) is more likely. This possibility has the additional advantage that it does not require generation of a new neural module. Rather, it could simply result from partial alterations to existing pathways for classical fear responses. Once this happened, selective inattention to death risk would also become possible.

- Notably, general reality denial and suppression of mortality fears are not by themselves positive features for an individual member of a species because of the danger of inappropriately risky behavior. Meanwhile, we have already said that acquiring a full ToM and understanding mortality would also be a negative feature at first. *But the simultaneous occurrence of these two negative attributes in the same individuals would cancel each other out*, allowing those developing a full ToM to survive and propagate by denying mortality risk.

- Once this combination became established in a small population, individuals with both abilities would have many positive benefits. For one thing, they would have all the benefits of full ToM, including the ability to understand and relate to each other—and they would also likely eschew mating with others who did not. In addition, reality denial also allows the emergence of optimism and overconfidence, which at a reasonable level can have many benefits and advantages to the individual and to the species.

- Thus the Mind over Reality theory posits that humans are the only extant species with full ToM because we are the only ones who crossed the evolutionary psychological barrier of living with the knowledge of our mortality by simultaneously acquiring the mental ability to deny reality.

- While it is currently not possible to falsify this hypothesis, it appears consistent with all available information. It can also explain a wide range of uniquely human features, such as risky behavior, the optimism bias, depressive realism, suicide, religiosity, existential angst, bravery, empathy, indirect reciprocity, and so on. And while there are many facts that fit with this theory, none appears as yet to directly militate against it.

- It is of course possible that full ToM appeared somewhat more gradually, coevolving step-by-step over time with denial of reality in the founder population of behaviorally modern humans. However, given the negative consequences and relative instability of each individual state, this coevolution could not have gone on for a very long time.

- Regardless of how the individual steps occurred, there should be an underlying neural basis for this unusual combination of cognitive and psychological features. In this regard it is interesting that the regions of the brain implicated in human ToM (the frontal cortex and its adjacent structures) are also the ones generally involved in executive control of both fear and optimism via interactions with the amygdala. Further studies could address the possibility that emergence of the neural substrate of a full ToM may have incidentally and partially altered pathways from the frontal cortex to the amygdala that mediate normal fear responses and conditioning, perhaps allowing a blunting of such responses. Meanwhile, it is intriguing that these brain areas are unusually modified in humans and are affected in some relevant mental disease states.

- Finally, taking all available information into consideration, the most likely timing of the proposed evolutionary transition would have been just prior to the emergence of behaviorally modern humans in Africa, currently estimated at around one hundred thousand years ago. Since the original effective population size of all living humans is estimated to be around five to ten thousand, the Mind over Reality transition that occurred first in a few individuals must have initially spread to others via breeding during this early phase. Once established, this population with a potent combination of novel mental abilities would then have gradually spread over the planet, eventually replacing other existing humanlike subspecies, with whom there was only limited and ineffective interbreeding.

ACKNOWLEDGMENTS

This book had a most unusual genesis, originating in the ideas of an individual who transferred them to me during a chance encounter and then died suddenly. For his intellectual gift, I am deeply indebted to Danny Brower, who chose to open up his fertile mind to me on the first and only occasion I met him. Of course, if Mani Ramaswami had not invited me to speak at the University of Arizona in 2005, this chance meeting would never have occurred. Following my realization that Danny had passed on, I was lucky to notice a dedication to him in an article by Sean Carroll, who then confirmed to me that Danny had published nothing about his ideas prior to his death. Pascal Gagneux, Sarah Hrdy, Kristen Hawkes, and Terry Sejnowski then listened to my early ramblings about Danny's idea and provided some important and critical feedback. It was Pascal (my resident expert for all questions related to evolution and primatology) who particularly encouraged me by his receptiveness to the underlying concept. Philip Campbell (chief editor of *Nature*) was very kind to then consider and publish my letter to the editor, which laid out my expansion of Danny's original theory. The story might have ended there if Danny's widow, Sharon Brower, had not heard about the *Nature* letter and contacted me. Sharon then provided me with Danny's draft manuscript and eventually encouraged me to finish his work. As I write this, I have yet to meet Sharon in person. Yet we have been able to agree on just about everything related to this book simply by talking on the phone (human theory of mind in full action!), and for that I am deeply grateful.

Along the way, a discussion with Randy Nesse provided valuable feedback from his unique perspectives on evolution and psychiatry. But by that stage in the process I was left with much uncertainty as to how to do justice to the project while still preserving Danny's original contributions. Critical feedback regarding my early attempts from Carl Zimmer, Chris Wills, Danny Povinelli, Francisco Ayala, Jim Handelman, Pascal Gagneux, Pat Churchland, and V. S. Ramachandran was extremely useful, helping me realize the need to take a more systematic and comprehensive approach to the subject. Danny's colleagues Ted Weinert, Peter Lawrence, and Mani Ramaswami also provided much valuable feedback and encouragement at this stage.

But despite all this intellectual support, I still could not have undertaken this project if it were not for the more than fifteen years I spent in the self-education program on human origins I created for myself at the La Jolla Group for Explaining the Origin of Humans (LOH), which morphed into the Project for Explaining the Origin of Humans (POH) and eventually into the Center for Academic Research and Training in Anthropogeny (CARTA). This improbable enterprise would also not have been possible without the support and encouragement of Rusty Gage (who first prompted me to start the LOH) and my other advisers—David Perlmutter, Kurt Benirschke, Jim Moore, Margaret Schoeninger, and Pascal Gagneux, who encouraged me to keep this improbable venture going. Many other members of LOH/POH/CARTA also helped educate me over the years on many aspects of anthropogeny. In addition, the more than three hundred talks relevant to anthropogeny that I have heard presented by

CARTA members over the course of more than fifteen years helped a great deal (many of the more recent ones can be found for public viewing at the CARTA website). And, of course, none of this could have happened if Jim Handelman and the board of the G. Harold and Leila Y. Mathers Foundation of New York had not provided long-term support for the LOH/POH/CARTA effort. Special mention should also go to Peter Preuss for supporting the very first LOH meeting and to UC San Diego and the Salk Institute for Biological Studies for their continued support in providing venues for our meetings.

While I was musing over what to do next, two previously published authors of popular books (Sanjay Nigam and Abraham Verghese) also gave me cogent advice and input. It was Abraham who suggested that I contact his literary agent, Mary Evans. I am deeply grateful to Mary for recognizing the potential of this "diamond in the rough," taking on the project enthusiastically, and following through vigorously to ensure that I produced a book proposal acceptable to a publisher and a manuscript targeted at an educated lay audience. Mary also spent much effort on persuading me to abandon my quest to preserve all Danny's original prose, assuring me that writing in a single voice was the best way to honor Danny's memory (something I finally brought myself to do, with Sharon Brower's agreement). Having then moved to the next draft, I was fortunate that many of my and Danny's friends and colleagues stepped in to provide valuable information, ideas, and criticisms. Having input from such a remarkable group of intelligent academics and laypeople really helped make the book more comprehensive, rigorous, and accessible. In addition to all those already mentioned, particular thanks should go to Alok Kalia, Claudia

Dunaway, Dan Geschwind, Derek Denton, Faye Flam, Ingrid Benirschke-Perkins, Jim Moore, Jon Cohen, Lori Marino, Philip Ninan, and Sarah Greene, all of whom provided extremely valuable feedback at various stages. I am also indebted to two individuals who tolerated and responded to a barrage of e-mails from me about specific topics: Tali Sharot (regarding the optimism bias) and Veerabhadran Ramanathan (regarding climate change). In-depth reads by Krish Sathian, Stephen Davies, and Pascal Gagneux played a very useful role before the manuscript was finalized. Special mention should also go to my publisher and editor, Cary Goldstein, who helped enormously all along the way, critiquing and reorganizing the text based on his extensive experience and helping me shape the message so that it would be more accessible.

Even at this stage I was in a quandary, because the manuscript still appeared to be in a mixture of two voices and hence somewhat difficult to read. But as I edited further, I was concerned that Danny's own voice was disappearing. Supportive assurances from Sharon Brower and cogent editorial suggestions from Cary Goldstein helped me get past this hurdle and power down the final stretch. Cary also assured Sharon and me that Danny's original draft manuscript would always be available to the interested reader via the publisher's website. A final round of diligent and insightful copyediting by Barbara Clark did much to eliminate unclear passages and optimize the final product. And Libby Burton of Twelve was always there, to promptly answer my numerous queries and to smooth the way for everything.

This book considers a vast swath of topics in numerous disciplines. Thus I am also deeply indebted to the many scientists, scholars, and published authors whose work and writings I cite, even while apologizing profusely for any citations that are missed, incomplete, or not entirely accurate. Given the centuries of human thought and writing on these diverse subjects, it is likely that some of the ideas and phrases I have come up with are not as original as I think they are. As a general resource I have also turned to the Internet, especially Wikipedia, Google Scholar, and TED.com (these remarkable and free sources of information provide unique portals to the world of knowledge available on the Web). On a broader note, I am grateful to those of my UC San Diego colleagues who tolerated my periodic mental withdrawals from my regular roles at the university and to the CARTA staff team, who protected my time by ensuring that the organization always ran smoothly and efficiently. Special mention must go to my assistant, Melanie Nieze, who not only transcribed many dictations and provided her own feedback but also helped keep my life otherwise in order during most of this arduous process (Susan Korosy and Joann Kim helped at the final stages). I also thank all the members of my own research group, who listened to my ramblings and musings and gave me valuable feedback.

Last but not least, I have to thank my own family, particularly my late grandfather Pothan Joseph, whose remarkable prowess as a journalist and editor inspired me to think one day of being a writer; my late father, Mathew Varki, who insisted on excellence in everything; my mother, Anna Varki, who ensured that I was exposed to English literature as I was growing up; my daughter, Sarah, who gave me pithy advice in the early stages of this project; and of course my long-suffering spouse and lifetime companion, Nissi, without whose unconditional love and caring none of this could have ever happened.

Acknowledgments

My coauthor is obviously not here to provide his contributions to this section. I asked Sharon Brower if there were any specific acknowledgments that Danny might have wanted to make, and she responded as follows: "Here is what I know and remember. As I recall, it started with Gio Bosco organizing the conference, and Mani Ramaswami inviting you to speak at it. But even before that, Danny had asked Sean Carroll, Mary Kaye Edwards, Ted Weinert, and Bruce Hubbard to read and comment on a draft version of his manuscript. Tom Bunch is the one who sent me your *Nature* article (without which none of this may have happened). Roy Parker, Ted Weinert, Mani Ramaswami, Murray Brilliant, Sean Carroll, and Peter Lawrence all spoke to Danny about his proposal, and he highly valued all of their opinions (and their friendship). There is huge risk that I have not mentioned everyone who was involved. I'm sure that Danny talked to quite a few people, and I have no idea who, if any, influenced his thinking. As I can't possibly know everything that he said and everyone that he spoke to, I offer here a profuse apology and a blanket thank you to those who were inadvertently left out."

NOTES

Introduction: An Improbable but True Story

1   DL Brower, "Platelets with Wings: The Maturation of Drosophila Integrin Biology," *Current Opinion in Cell Biology* 15 (2003): 607–13; DL Brower, SM Brower, DC Hayward, and EE Ball, "Molecular Evolution of Integrins: Genes Encoding Integrin Beta Subunits from a Coral and a Sponge," *Proceedings of the National Academy of Sciences of the United States of America* 94 (1997): 9182–87; BP James, TA Bunch, S Krishnamoorthy, LA Perkins, and DL Brower, "Nuclear Localization of the Erk Map Kinase Mediated by Drosophila Alphaps2Betaps Integrin and Importin-7," *Molecular Biology of the Cell* 18 (2007): 4190–99.

2   There are numerous excellent books on evolution for the general reader. Here are some examples: Richard Dawkins, *The Selfish Gene*, 30th anniversary ed. (Oxford University Press, 2006); Mark Ridley, *Evolution* (Oxford University Press, 2004); Sean B. Carroll, *Endless Forms Most Beautiful: The New Science of Evo Devo and the Making of the Animal Kingdom* (W. W. Norton, 2005); Carl Zimmer, *Evolution: The Triumph of an Idea* (HarperPerennial, 2006); David Sloan Wilson, *Evolution for Everyone: How Darwin's Theory Can Change the Way We Think About Our Lives* (Delacorte Press, 2007); Franz M. Wuketits and Francisco J. Ayala, eds., *Handbook of Evolution: The Evolution of Living Systems (Including Hominids)* (Wiley-Blackwell, 2005).

3   A Varki, RD Cummings, JD Esko, HH Freeze, P Stanley, CR Bertozzi, GW Hart, and ME Etzler, *Essentials of Glycobiology*, 2nd ed. (Cold Spring Harbor Laboratory Press, 2009); A Varki, "Evolutionary Forces Shaping the Golgi Glycosylation Machinery: Why

Cell Surface Glycans Are Universal to Living Cells," *Cold Spring Harbor Perspectives in Biology* 3 (2011), DOI: 10.1101/cshperspect .a005462.

4   National Research Council, *Transforming Glycoscience: A Roadmap for the Future* (National Academies Press, 2012). I was a member of the committee that generated this report.

5   No amount of prior experience or education can prepare a first-time parent for the sheer wonder of watching a helpless newborn gradually develop into a fully mature human.

6   M Goodman and KN Sterner, "Colloquium Paper: Phylogenomic Evidence of Adaptive Evolution in the Ancestry of Humans," *Proceedings of the National Academy of Sciences of the United States of America* 107, suppl. 2 (2010): 8918–23. There are many other relevant publications, beginning with Charles Darwin and Thomas Huxley. I cite the late Morris Goodman because he was one of the first to provide molecular evidence for this close evolutionary relationship.

7   HH Chou, H Takematsu, S Diaz, J Iber, E Nickerson, KL Wright, EA Muchmore, DL Nelson, ST Warren, and A Varki, "A Mutation in Human CMP-Sialic Acid Hydroxylase Occurred after the Homo-Pan Divergence," *Proceedings of the National Academy of Sciences of the United States of America* 95 (1998): 11751–56; HH Chou, T Hayakawa, S Diaz, M Krings, E Indriati, M Leakey, S Pääbo, Y Satta, N Takahata, and A Varki, "Inactivation of CMP-N-Acetylneuraminic Acid Hydroxylase Occurred Prior to Brain Expansion during Human Evolution," *Proceedings of the National Academy of Sciences of the United States of America* 99 (2002): 11736–41.

8   A Varki, "Colloquium Paper: Uniquely Human Evolution of Sialic Acid Genetics and Biology," *Proceedings of the National Academy of Sciences of the United States of America* 107, suppl. 2 (2010): 8939–46; NM Varki, E Strobert, EJ Dick Jr., K Benirschke, and A Varki, "Biomedical Differences between Human and Nonhuman Hominids: Potential Roles for Uniquely Human Aspects of Sialic Acid Biology," *Annual Review of Pathology* 6 (2011): 365–93.

9   In looking for a single term to best encapsulate the systematic exploration of human origins, I came across the word *anthropogeny*,

which was in use as early as 1839 (Robert Hooper's *Medical Dictionary* defined it then as "the study of the generation of man"). And it is still listed in the *Oxford English Dictionary*. But while biologist Ernst Haeckel used the term in the late 1880s and defined it as "the history of the evolution of man," it seems to have been lost to usage after the 1930s. I decided to resurrect the term in naming the Center for Academic Research and Training in Anthropogeny (CARTA).

10   What began as the La Jolla Group for Explaining the Origin of Humans (LOH) morphed into the Project for Explaining the Origin of Humans (POH) and, eventually, into CARTA. Please see the acknowledgments for the names of some key individuals who were involved in the process.

11   The G. Harold and Leila Y. Mathers Foundation of New York is one of those very quiet philanthropic organizations that does enormous good but seeks no attention or credit whatsoever. While primarily focused on funding basic research in the life sciences, the foundation has also chosen to support this transdisciplinary effort. I am forever grateful to the director, Jim Handelman, and his foundation board for their generous support of this improbable venture.

12   Main CARTA website: http://carta.anthropogeny.org/.

13   The complete CARTA mission statement is as follows: "Use all rational and ethical approaches to seek all verifiable facts from all relevant disciplines to explore and explain the origins of the human phenomenon, while minimizing complex organizational structures and hierarchies, and avoiding unnecessary paperwork and bureaucracy. In the process, raise awareness and understanding of the study of human origins within the academic community and the public at large."

14   As we will discuss later, the term *self-awareness* has many meanings. I use it here to indicate the likelihood of individuals of a species understanding and recognizing their own personhood.

15   SB Carroll, "Evo-Devo and an Expanding Evolutionary Synthesis: A Genetic Theory of Morphological Evolution," *Cell* 134 (2008): 25–36.

16   P Gagneux, JJ Moore, and A Varki, "The Ethics of Research on
      Great Apes," *Nature* 437 (2005): 27–29.

17   Having been raised as an Indian Christian, I did not know much
      about Indian Vedic texts—the Mahabharata story was pointed out
      to me by Kalyan Banda, a research fellow in my lab.

18   Ernest Becker, *The Denial of Death* (Free Press, 1973).

19   A Varki, "Human Uniqueness and the Denial of Death," *Nature*
      460 (2009): 684.

20   A Rosenblatt, J Greenberg, S Solomon, T Pyszczynski, and D
      Lyon, "Evidence for Terror Management Theory: I. The Effects of
      Mortality Salience on Reactions to Those Who Violate or Uphold
      Cultural Values," *Journal of Personality and Social Psychology* 57
      (1989): 681–90; S Solomon, J Greenberg, and T Pyszczynski, "A
      Terror Management Theory of Social Behavior: The Psychologi-
      cal Functions of Self-Esteem and Cultural Worldviews," *Advances
      in Experimental Social Psychology* 24 (1991): 93–159; J Green-
      berg, S Solomon, and T Pyszczynski, "Terror Management Theory
      of Self-Esteem and Cultural Worldviews: Empirical Assessments
      and Conceptual Refinements," *Advances in Experimental Social
      Psychology* 29 (1997): 61–139; T Pyszczynski, J Greenberg, and
      S Solomon, "A Dual-Process Model of Defense against Conscious
      and Unconscious Death-Related Thoughts: An Extension of Terror
      Management Theory," *Psychological Review* 106 (1999): 835–45.

21   This introduction is a blending of my own first draft prologue with
      Danny's original introductory chapter. Sharon Brower explained
      that Danny deliberately wrote his first two drafts without consult-
      ing many experts, to first just get his own unbiased observations
      down on paper. I am truly fortunate to have recruited interested
      experts and laypeople (including Danny's friends and colleagues)
      who were willing and able to provide comments from a reader's
      perspective (please see the listing of names in the acknowledg-
      ments).

22   Tali Sharot, *The Optimism Bias: A Tour of the Irrationally Positive
      Brain* (Pantheon, 2011).

23   The CARTA Library of Anthropogeny (http://carta.anthropogeny
      .org/libraries/anthropogeny/biblio) is an online chronological list-
      ing of a selection of the many published books relevant to ex-

ploring anthropogeny. Some of these books (and many others that have recently appeared) are cited at various points throughout this volume.

24   In the course of my work with CARTA and its precursor organizations, I have had the privilege of listening to more than three hundred lectures by experts on all aspects of anthropogeny. So while I will not present an extensive bibliography with citations to primary literature, as an expert author might do, I am reasonably confident about the basic concepts presented. I will also set forth unknowns and caveats throughout.

25   Definitions of the word *denial* from *The American Heritage Dictionary of the English Language*, 4th ed.:

1. A refusal to comply with or satisfy a request.

2a. A refusal to grant the truth of a statement or allegation; a contradiction.

b. *Law* The opposing by a defendant of an allegation of the plaintiff.

3a. A refusal to accept or believe something, such as a doctrine or belief.

b. *Psychology* An unconscious defense mechanism characterized by refusal to acknowledge painful realities, thoughts, or feelings.

4. The act of disowning or disavowing; repudiation.

5. Abstinence; self-denial.

26   I have borrowed this phrase from Charles Darwin, who used it to summarize his approach to a novel theory (descent by evolution via natural selection) that seemed to explain everything he knew but that he could not prove by any experiment. The phrase is also within the title of a book by the great evolutionary biologist Ernst Mayr: *One Long Argument: Charles Darwin and the Genesis of Modern Evolutionary Thought* (Harvard University Press, 1993).

27   These were the words of Darwin's friend Thomas Huxley. A good idea in science will eventually be pursued to the point of either being overturned or exalted to the status of "fact." So, as eventually happened with Darwin, we hope that other scientists and thinkers will come up with ways to more precisely test the theory we will present.

Chapter 1. Where Did We Come From, and How Did We Get Here?

1   http://carta.anthropogeny.org/symposia. CARTA typically orga-
      nizes three public symposia a year. These free meetings bring
      together scientists eminent in their respective fields to address
      some particular aspect of human origins. Presentations are di-
      rected to scientists in other fields and to an educated lay audience,
      and attempt to minimize the use of jargon while highlighting clear
      and simple messages. The location of the symposia rotates be-
      tween the Salk Institute, the UC San Diego campus, and the UC
      San Diego School of Medicine. Each presentation is recorded on
      video and later broadcast by UCTV as well as on iTunes U and
      YouTube before being archived on the CARTA and UCTV web-
      sites.

2   Updated from A Varki, "Nothing in Medicine Makes Sense, Except
      in the Light of Evolution," *Journal of Molecular Medicine* 90
      (2012): 481–94, used as a handout for an introductory lecture to
      medical students. The article is based on knowledge from numer-
      ous sources, including many of the books cited in this volume. For
      another brief summary, see Bernard Wood, *Human Evolution: A
      Brief Insight* (Sterling, 2011).

3   Jared Diamond, *The Third Chimpanzee: The Evolution and Future
      of the Human Animal* (HarperPerennial, 1993).

4   Ann Gibbons, *The First Human: The Race to Discover Our Earliest
      Ancestors* (Anchor Books, 2006).

5   TD White, B Asfaw, Y Beyene, Y Haile-Selassie, CO Lovejoy,
      G Suwa, and G WoldeGabriel, "Ardipithecus Ramidus and the Pa-
      leobiology of Early Hominids," *Science* 326 (2009): 75–86; A Gib-
      bons, "Breakthrough of the Year: Ardipithecus Ramidus," *Science*
      326 (2009): 1598–99.

6   Donald Johanson and Blake Edgar, *From Lucy to Language*, re-
      vised, updated, and expanded ed. (Simon and Schuster, 2006);
      Donald Johanson and Kate Wong, *Lucy's Legacy: The Quest for
      Human Origins* (Broadway, 2010).

7   Richard E. Leakey and Roger Lewin, *Origins Reconsidered: In
      Search of What Makes Us Human* (Anchor, 1993).

8   DM Bramble and DE Lieberman, "Endurance Running and the Evolution of Homo," *Nature* 432 (2004): 345–52.

9   Clive Finlayson, *The Humans Who Went Extinct: Why Neanderthals Died Out and We Survived* (Oxford University Press, 2010).

10   Chris Stringer, *Lone Survivors: How We Came to Be the Only Humans on Earth* (Henry Holt, 2012).

11   The exact timing of the origin of behaviorally modern humans in Africa is constantly being revisited and refined. I will use "around one hundred thousand years ago" here and throughout as a rough approximation.

12   Given the very short time since we emerged from Africa and subsequent backward and forward migrations and admixtures, there is no biological basis for the concept of "race." Rather, there seem to be "clines," with small differences found between human groups as one goes in geographically different directions.

13   Darwin got there a bit ahead, but sat on his idea for years, apparently to avoid the controversy that would surely erupt when he published it. But Wallace wrote Darwin a letter asking for his opinion about a theory of evolution he had come up with. Darwin was aghast that someone else had come up with the very same idea he was sitting on. But he did the right thing and arranged to co-publish with Wallace.

14   Both men were influenced by Thomas Malthus, who noted that since a human population increases exponentially and food availability does not, famine will restrict population growth. Darwin and Wallace applied this idea to all living things, realizing that for an expanding population, resources become limiting. By "resources," they meant everything an organism needs to survive and reproduce, such as food and shelter—and, for sexually reproducing organisms, mates. Once a component necessary for optimal population expansion becomes scarce, there will be competition for it—between different types of organisms for a single resource or between different members of a single species. Some members of a species are bigger, faster, stronger, or smarter. Others can better survive hard times because of variations in metabolism. Yet others compete better for mates. Better competitors leave behind

more progeny. Darwin and Wallace understood that individual differences are based on inheritable "traits" (features) that are passed on to progeny. It was decades before others showed what "genes" are, but they had the basic idea spot-on.

15   Hybrids such as a mule result from mating a female horse and male donkey. Some rare hybrids are fertile, such as the occasional female liger (resulting from the mating of a male lion and female tiger). Male ligers are always sterile, however, and these hybrids result from artificial conditions in captivity.

16   Darwin was impressed by different types of Galápagos finches, which had evolved independently on separate islands. A difference in food seed hardness had selected for different beak shapes and sizes. The birds were similar enough to suggest a common ancestry but different in ways that could be explained by environmental variations. Wallace later noted a more dramatic case— a geographic feature now called Wallace's line, which is a deep ocean channel that separates the tigers and monkeys of Indonesia from the kangaroos and koala bears of Australia.

17   For example, differences in chromosomes (the highest level of DNA organization in a cell) can limit the generation of viable or fertile offspring. Such biological barriers to "gene flow" can be every bit as effective as geological barriers. Behavioral change can also lead to genetic isolation if it alters mating success. Seasonal timing can also matter—if a mutation in a tree causes it to flower at a different time of year from its cohort, then there will be little or no chance of cross-pollination, and so on.

18   E Joly, "The Existence of Species Rests on a Metastable Equilibrium between Inbreeding and Outbreeding: An Essay on the Close Relationship between Speciation, Inbreeding, and Recessive Mutations," *Biology Direct* 6 (2011): 62. In this model, speciation is driven by several advantages of inbreeding, mainly advantageous recessive phenotypes that could only be retained by inbreeding. Reproductive barriers would thus not arise as secondary consequences of divergent evolution in populations isolated from one another but rather under the direct selective pressure of ancestral stocks.

19   D Ghaderi, SA Springer, F Ma, M Cohen, P Secrest, RE Taylor,

A Varki, and P Gagneux, "Sexual Selection by Female Immunity against Paternal Antigens Can Fix Loss of Function Alleles," *Proceedings of the National Academy of Sciences of the United States of America* 108 (2011): 17743–48.

20   It is important to note here that no one can say with certainty how life originated. Molecules needed for life could have arisen from components of primal earth, but it is also possible that molecules (or even living forms) could have arrived from meteorites or comets. We do, however, have a pretty good idea about the evolutionary history of species (especially big species, such as plants and animals) over the past few hundred million years. Of course, anyone who asserts that a divine creator did not initially start this ball rolling is basing that belief on the same level of faith as those who make the counterargument. There is just no evidence of divine interference after evolution was set in motion. Indeed, neither Danny nor I have or had anything against faith and spirituality. These uniquely human mind-sets have done much to comfort and support humans and are related to the theme of this book. But faith and spirituality should never be confused with science, which deals with what can be verified by objective observations (in other words, "reality"). It is absurd to promote an "intelligent design science" based on ignorance, and difficult to take seriously any "creation scientist" who asserts that just because he or she can't explain something it is therefore unexplainable. Indeed, it is ironic that the argument for an intelligent creator is based on our ignorance of the way in which events in evolution might have occurred. Since this ignorance is continuously decreasing, our concept of a creator must also be constantly evolving.

21   Such as dinosaurs with feathers—the ancestors of present-day birds. See M Balter, "Paleontology: Flying Dinos and Baby Birds Offer New Clues about How Avians Took Wing," *Science* 338 (2012): 591–92.

22   This has left a striking footprint—similarities between organisms. Structures similar in two organisms are said to be "homologous" when they result from common ancestry. At the molecular level, homologies are nothing short of astounding. All living things on

the planet use the same molecules to carry genetic instructions (DNA and RNA), and all translate this information in much the same way. For example, the metabolic reactions that extract energy are nearly identical in all animals, and similar to those in bacteria. At this biochemical level, we also see examples of how something doing one job is modified during evolution to take on a different function.

23 Some examples result from human activities, which drives "unnatural" selection. Textbooks describe the case of a species of moth in England (*Biston betularia*). In the nineteenth century, extensive coal burning darkened the roosting places of the moths, and there was rapid selection for a darker form, which blended better with the darker background. While the dark moth form became more common in sooty urban areas, lighter moths persisted in the countryside. Birds were more likely to spot and lunch on moths whose color contrasted with the background. As pollution controls improved and trees lightened up, so did the overall color of the moths. It is likely that bird predation was a major selective force—an assertion that hasn't been proved—but the underlying concept is not considered controversial, except among some evolution doubters. Indeed, elucidation of the molecular basis for color selection may be close at hand. See AE van't Hof, N Edmonds, M Dalikova, F Marec, and IJ Saccheri, "Industrial Melanism in British Peppered Moths Has a Singular and Recent Mutational Origin," *Science* 332 (2011): 958–60.

24 Your immune system has evolved to identify and attack foreign substances. This is true for invading cells, such as bacteria, as well as for your own cells, especially when they have been infected by a virus. The immune system has also evolved to "remember" foreign things it has seen before and attack them with special vigor on return. This "immunological memory" is why you don't get many infectious diseases more than once and underlies the way in which vaccination protects you from infections (arguably one of the greatest achievements of modern medicine and public health).

25 Many antibiotics can cure bacterial infections. But a subset of bacteria is resistant to these drugs to varying degrees, and the more antibiotics we take the more we select for these resistant bugs.

Why is there a genetic foundation for antibiotic resistance in a natural population of bacteria if humans have only been using antibiotics recently? Because most antibiotics are based on natural compounds that other organisms have been using to fight bacteria for eons.

26   Although they are not living organisms in the true sense, viruses are genetic entities with DNA or RNA genes that invade and hijack your own cells to reproduce themselves. Viruses typically mutate their genes at a high rate, allowing them to keep one step ahead of the immune system. You have to get a new flu shot each year because a slightly different virus strain is expected to be common each season. Occasionally, a flu strain in another animal evolves to infect humans. When these strains also evolve genetic variations necessary for transmission from human to human, a completely new strain enters the population, which is different from any our immune systems have seen before. These evolutionary jumps into humans seem to be responsible for the great influenza pandemics.

27   HIV (the causative agent of acquired immunodeficiency syndrome—AIDS) is an RNA virus that mutates very quickly. It evolves drug resistance readily, so no single agent is successful. Viral genomes circulating during full-blown AIDS infection are even different from the virus that initially infected the very same person.

28   These are different populations in nature that are still potentially or actually interbreeding, albeit at a reduced rate. Are these different species? Where can one draw a species boundary? To give an example resulting from unnatural selection, all domestic dogs are one species. But if all dogs were to disappear except for tiny Chihuahuas and massive Great Danes who were then let loose in the wild, there would likely be no interbreeding between these populations for obvious reasons of physical incompatibility. Would they then be separate species? They certainly would evolve independently and might eventually become incompatible genetically, not just physically.

29   RE Green, J Krause, AW Briggs, T Maricic, U Stenzel, M Kircher, N Patterson, H Li, W Zhai, MH Fritz, NF Hansen, EY Durand, AS Malaspinas, JD Jensen, T Marques-Bonet, C Alkan, K Prufer,

M Meyer, HA Burbano, JM Good, R Schultz, A Aximu-Petri, A Butthof, B Hober, B Hoffner, M Siegemund, A Weihmann, C Nusbaum, ES Lander, C Russ, N Novod, J Affourtit, M Egholm, C Verna, P Rudan, D Brajkovic, Z Kucan, I Gusic, VB Doronichev, LV Golovanova, C Lalueza-Fox, M de la Rasilla, J Fortea, A Rosas, RW Schmitz, PL Johnson, EE Eichler, D Falush, E Birney, JC Mullikin, M Slatkin, R Nielsen, J Kelso, M Lachmann, D Reich, and S Pääbo, "A Draft Sequence of the Neandertal Genome," *Science* 328 (2010): 710–22; S Sankararaman, N Patterson, H Li, S Pääbo, and D Reich, "The Date of Interbreeding between Neandertals and Modern Humans," *PLoS Genetics* 8 (2012): e1002947.

30   M Meyer, M Kircher, MT Gansauge, H Li, F Racimo, S Mallick, JG Schraiber, F Jay, K Prufer, C de Filippo, PH Sudmant, C Alkan, Q Fu, R Do, N Rohland, A Tandon, M Siebauer, RE Green, K Bryc, AW Briggs, U Stenzel, J Dabney, J Shendure, J Kitzman, MF Hammer, MV Shunkov, AP Derevianko, N Patterson, AM Andres, EE Eichler, M Slatkin, D Reich, J Kelso, and S Pääbo, "A High-Coverage Genome Sequence from an Archaic Denisovan Individual," *Science* 338 (2012): 222–26.

31   Your genes are composed of DNA organized in chromosomes, which are duplicated when a cell divides. Each gene contains information for making a specific protein. Proteins are made of amino acids, and fold up to make little molecular machines. These proteins control chemical reactions and comprise parts of your cells. Some proteins digest and extract energy from the food you eat. Others are molecular motors, responsible for things like muscle movements. Lipids (fats) and sugar chains (glycans) make up the rest of the key components of your cells. Genes are like recipes in a cookbook, and it is essential that they be turned on and off properly so that they'll make the right proteins in the right quantity at the right time and place. There are only about twenty to twenty-five thousand unique protein-coding genes in a human genome, so most of our DNA does not actually contain information for making proteins. Some of this extra DNA was known to be important for controlling when a gene will be used, but most was until recently thought to be doing nothing relevant. But it has recently been discovered that much of this so-called junk DNA

has functions that we did not previously know about. For example, there are many DNA-encoded "non-coding RNAs" that have important functions, such as switching genes off and on. In fact, recent data indicate that more than 80 percent of your genome may be active.

32   Matt Ridley, *Nature via Nurture: Genes, Experience, and What Makes Us Human* (HarperCollins, 2003).

33   Evelyn Fox Keller, *The Mirage of a Space between Nature and Nurture* (Duke University Press, 2010).

34   Diabetes provides an interesting example of the interplay between genes and environment. In his original manuscript, Danny had a discussion about this matter. It is not reproduced here due to lack of space. Also not mentioned here is the related science of epigenetics, which shows that transgenerational inheritance does not have to be based on genes alone but can also be affected by special marks on DNA and/or the proteins that surround DNA on chromosomes.

35   Note that natural selection is only peripherally concerned with genes promoting your long-term survival. Certainly if you die before reaching reproductive age, your fitness is zero. However, once you have had offspring, the impact of selection is limited. In humans and some other animals, however, the parents, and even grandparents, contribute critical nurturing and learning benefits to their offspring.

36   Not necessarily absolute zero. He might be able to slip in for a mating opportunity on the sly—for example, Robert Sapolsky and his colleagues report that some male orangutans take an alternative approach to reproductive success. A small and weak-looking male will surreptitiously have sex (often forced) with females when the large strong male is not around: AN Maggioncalda, NM Czekala, and RM Sapolsky, "Male Orangutan Subadulthood: A New Twist on the Relationship between Chronic Stress and Developmental Arrest," *American Journal of Physical Anthropology* 118 (2002): 25–32.

37   As an aside, when it comes to reproductive fitness, the human species may be a peculiar anomaly. There is a thriving industry built around infertility, ranging from magical or quack "cures"

to sophisticated in vitro fertilization to surrogate pregnancies. In some cases the underlying cause is evident. On the male side, there can be low sperm numbers or poor sperm quality. On the female side, there are common issues such as polycystic ovarian disease and endometriosis, diseases that appear to be relatively human-specific. Another negative factor is gonorrhea, a sexually transmitted disease unique to humans. And even among women who become pregnant it is very common to lose the fetus either early on due to chromosomal problems or later on due to a common problem called preeclampsia—yet another apparently human-specific disease. Finally, the process of human labor is fraught with dangers for the mother and baby because of the unusually large size of the human head, which butts up against the narrow human birth canal. Given our repeated mantra that reproductive fitness is the key to evolutionary success, the high rate of infertility and fetal loss in humans seems almost paradoxical. Did factors beyond the conventional principles of evolution contribute to the success of the human species?

38 Examples are rife: Just a few are cited in JS Olshansky, BA Carnes, and RN Butler, "If Humans Were Built to Last," *Scientific American* 284 (2001): 50–55.

39 Updated from A Varki, "Nothing in Medicine Makes Sense, Except in the Light of Evolution," *Journal of Molecular Medicine* 90 (2012): 481–94. I provided this proof as a handout for an introductory lecture to medical students.

A Proof of Evolution by First Principles
- Over time, humans in many societies have classified living organisms into groups such as animals, plants, and fungi (and, more recently, microbes) based on our observations of similarities and differences.
- Some humans further divided such groups into subgroups and into subgroups of subgroups, again based on observational criteria (initially, external appearance; later, internal features).
- When a particular subgroup showed great similarities between individuals and they bred and reproduced, giving rise to similar individuals in the next generation, we humans called such a subgroup a species.

- Much later, humans found that all known life forms require DNA as a "genetic code."
- It then emerged that the genetic-code lettering system of DNA is essentially the same in all life forms.
- The relatedness of DNA sequences from different species was then found to be almost exactly in line with the prior classification of species based on other observations. So in the "tree of life," living things appear related to one another by their DNA.
- At the level of populations, all species show individual variations in DNA, their bodies, and/or their behavior, which can be beneficial, neutral, or detrimental depending on the environmental conditions.
- Such variation is also common in the DNA of different individuals within natural populations.
- During reproduction, the DNA of one generation must be passed on to offspring. This process introduces variation into the progeny by less-than-perfect replication of DNA.
- Ongoing random mutations also introduce further variations and changes into DNA.
- Most species produce far more progeny than can possibly survive. What prevents natural populations from explosive expansion is that only a small fraction of all progeny survive and reproduce.
- Individuals who reproduce (pass their DNA on to progeny) are likely to be those whose variations were most beneficial and/or least detrimental under the circumstances (a nonrandom process).
- Thus populations inevitably change over time, as DNA variations beneficial to prevailing environments accumulate over generations. This process can be called natural selection, and it results in adaptations to environmental conditions.
- Many such adaptations eventually appear to be very exquisite designs, but on closer observation they are still imperfect and/or seem constructed in an illogical fashion.
- Mate choice can influence which individual's DNA is passed on. This is called sexual selection and can lead to astonishing

features of no apparent survival value to the individual, e.g., the male peacock's tail and the male moose's antlers.

• In addition to natural and sexual selection, the distribution of DNA variations and imperfections across populations can occur randomly, without selection. This is called neutral drift.

• Working together, natural selection, sexual selection, and neutral drift can lead to differences between populations that are eventually large enough to generate barriers to successful mating.

• Such "reproductive isolation" will eventually give rise to new species, as newly isolated populations accumulate independent DNA changes and cease to have any DNA exchange.

• Once a beneficial DNA change becomes important for survival and reproduction, any further changes tend to be detrimental and cannot be tolerated without losing the individual in whom the change occurs (and hence that individual's DNA). Thus some aspects of DNA remain "conserved" during passage to the next generation. This is called purifying selection.

Taken together, all this information *can only be explained by assuming that all life forms are related by a single genetic code and have diverged over time into the different species we see today* via processes such as natural selection, sexual selection, neutral drift, and purifying selection. The sum total of all these processes can be called biological evolution. It is today the *only possible fact-based explanation* for the existence of so many life forms on earth, with all their variations and imperfections. No other explanations come even close. Are you now convinced? If not, you may not be able to follow the rest of this book.

40   T Sejnowski and T Delbruck, "The Language of the Brain," *Scientific American* 307 (2012): 54–59.

41   There have been numerous interesting books written about the human brain and its quirks, viewed from many different perspectives. Here are some examples: John S. Allen, *The Lives of the Brain: Human Evolution and the Organ of Mind* (Belknap Press of Harvard University Press, 2009); Daniel Bor, *The Ravenous Brain: How the New Science of Consciousness Explains Our Insatiable Search for Meaning* (Basic Books, 2012); Dean Buono-

mano, *Brain Bugs: How the Brain's Flaws Shape Our Lives* (W. W. Norton, 2011); Rita Carter, *Exploring Consciousness* (University of California Press, 2002); Jean-Pierre Changeux, *Neuronal Man: The Biology of Mind* (Princeton University Press, 1997); Antonio R. Damasio, *Descartes' Error: Emotion, Reason, and the Human Brain* (Quill, 2000); Terrence Deacon, *The Symbolic Species: The Co-Evolution of Language and the Brain* (W. W. Norton, 1997); David Eagleman, *Incognito: The Secret Lives of the Brain* (Vintage, 2012); Gerald M. Edelman and Giulio Tononi, *A Universe of Consciousness: How Matter Becomes Imagination* (Basic Books, 2000); John F. Hoffecker, *Landscape of the Mind: Human Evolution and the Archaeology of Thought* (Columbia University Press, 2011); Bruce Hood, *The Self Illusion: How the Social Brain Creates Identity* (Oxford University Press, 2012); Donald W. Pfaff, *The Neuroscience of Fair Play: Why We (Usually) Follow the Golden Rule* (Dana Press, 2007); Steven Pinker, *The Language Instinct: How the Mind Creates Language* (HarperPerennial, 2000); Steven R. Quartz and Terrence J. Sejnowski, *Liars, Lovers, and Heroes: What the New Brain Science Reveals about How We Become Who We Are* (William Morrow, 2002); V. S. Ramachandran and Sandra Blakeslee, *Phantoms in the Brain: Probing the Mysteries of the Human Mind* (Quill, 1999); Terrence W. Deacon, *Incomplete Nature: How Mind Emerged from Matter* (W. W. Norton, 2011).

42　If you are a parent, you will have no trouble accepting this premise based on personal experience. It is clear that children have different personalities virtually from the day they were born, and in each of them it is easy to detect distinct aspects of one or the other parent. These differences cannot be blamed on different environments. And anyone who has dogs or cats knows they have individual personalities, which are also not much changed by their environment.

43　Jerome Kagan and Nancy Snidman, *The Long Shadow of Temperament* (Belknap Press of Harvard University Press, 2009).

44　Joseph E. LeDoux, *The Emotional Brain: The Mysterious Underpinnings of Emotional Life* (Simon and Schuster, 1998); Paul Ekman, *Emotions Inside Out: 130 Years after Darwin's* The Expression of the Emotions in Man and Animals, *Annals of the*

New York Academy of Sciences 1000 (2003); Daniel Goleman, *Emotional Intelligence: Why It Can Matter More Than IQ*, 10th anniversary ed. (Bantam Books, 2005).

45 A Varki, DH Geschwind, and EE Eichler, "Explaining Human Uniqueness: Genome Interactions with Environment, Behaviour, and Culture," *Nature Reviews Genetics* 9 (2008): 749–63. The second part of this review contains many ideas relevant to this book.

46 http://www.darwinawards.com/darwin/.

47 Biological evolution is much slower than cultural and social evolution. So our lives are changing at a rate that has far outpaced evolution in our genes. There is a mismatch between the environments in which we evolved and the novel ones that we have created for ourselves. Yet we still carry our biological evolutionary baggage and must use this genetic tool kit to adapt to modern life. This insight is going to be true not only for aspects of our physiology but also for the evolutionary component of our personalities. It can be tempting to dismiss the effect of selection on behavior, since our behaviors don't always seem to be well adapted to our lives. But the genetic component of our personalities was selected to flourish in an environment foreign to most of us today, and we changed the rules much faster than our genes can keep up with. If we realize that we were not evolved (by selection) to live in our current world, it becomes easier to understand that our behaviors will not always make sense. This becomes even truer when we realize that evolution's "goals" are not necessarily ours.

48 Colin Renfrew, *Prehistory: The Making of the Human Mind* (Modern Library, 2009).

49 M Pagel, "Evolution: Adapted to Culture," *Nature* 482 (2012): 297–99; Mark Pagel, *Wired for Culture: Origins of the Human Social Mind* (W. W. Norton, 2012).

50 LG Dean, RL Kendal, SJ Schapiro, B Thierry, and KN Laland, "Identification of the Social and Cognitive Processes Underlying Human Cumulative Culture," *Science* 335, no. 6072 (2012): 1114–18. See also the discussion by R Kurzban and HC Barrett in "Behavior: Origins of Cumulative Culture," *Science* 335 (2012): 1056–57.

51 Ian Tattersall, *Masters of the Planet: The Search for Our Human Origins* (Palgrave Macmillan, 2012).

52   William H. Calvin, *A Brief History of the Mind: From Apes to Intellect and Beyond* (Oxford University Press, 2005).

Chapter 2. Becoming Smarter Shouldn't Be Hard

1   There are numerous excellent books addressing the question of human uniqueness. Here are a few recent examples (others are cited throughout the book): Chris Stringer, *The Origin of Our Species* (Allen Lane, 2011); David P. Barash, *Homo Mysterious: Evolutionary Puzzles of Human Nature* (Oxford University Press, 2012); Kim Sterelny, *The Evolved Apprentice: How Evolution Made Humans Unique* (Bradford Books, 2012); Philip Lieberman, *The Unpredictable Species: What Makes Humans Unique* (Princeton University Press, 2013).

2   Jane Goodall, *My Life with the Chimpanzees*, revised ed. (Aladdin, 1996); Jane Goodall, *Through a Window: My Thirty Years with the Chimpanzees of Gombe*, 50th anniversary of Gombe ed. (Mariner Books, 2010).

3   Frans de Waal, *Our Inner Ape: A Leading Primatologist Explains Why We Are Who We Are* (Riverhead, 2006); Frans de Waal, *The Age of Empathy: Nature's Lessons for a Kinder Society* (Harmony Books, 2009).

4   William C. McGrew, *Chimpanzee Material Culture: Implications for Human Evolution* (Cambridge University Press, 1992).

5   Tetsuro Matsuzawa, Masaki Tomonaga, and Masayuki Tanaka, *Cognitive Development in Chimpanzees* (Springer, 2011).

6   Daniel J. Povinelli, *Folk Physics for Apes: The Chimpanzee's Theory of How the World Works* (Oxford University Press, 2003); Daniel J. Povinelli, *World without Weight: Perspectives on an Alien Mind* (Oxford University Press, 2012).

7   Michael Tomasello, *Origins of Human Communication* (MIT Press, 2008); Michael Tomasello, *Why We Cooperate* (MIT Press, 2009).

8   David Premack and Ann Premack, *Original Intelligence: Unlocking the Mystery of Who We Are* (McGraw-Hill, 2003).

9   Herbert S. Terrace and Janet Metcalfe, *The Missing Link in Cogni-*

tion: Origins of Self-Reflective Consciousness (Oxford University Press, 2005).

10   Jon Cohen, *Almost Chimpanzee: Searching for What Makes Us Human, in Rainforests, Labs, Sanctuaries, and Zoos* (Times Books, 2010).

11   Jeremy Taylor, *Not a Chimp: The Hunt to Find the Genes That Make Us Human* (Oxford University Press, 2010).

12   Raymond Tallis, *Aping Mankind: Neuromania, Darwinitis, and the Misrepresentation of Humanity* (Acumen Publishing, 2011).

13   Some examples are: Andrew Whiten and Richard W. Byrne, *Machiavellian Intelligence II: Extensions and Evaluations* (Cambridge University Press, 1997); Robin Dunbar, *Grooming, Gossip, and the Evolution of Language* (Harvard University Press, 1998); Robin Dunbar, *How Many Friends Does One Person Need? Dunbar's Number and Other Evolutionary Quirks* (Harvard University Press, 2010).

14   Matt Ridley, *The Red Queen: Sex and the Evolution of Human Nature* (HarperPerennial, 2003).

15   Stephen Davies, *The Artful Species: Aesthetics, Art, and Evolution* (Oxford University Press, 2013).

16   At a 2002 symposium on human evolution that I organized, Francis Crick observed that "there are no [absolute] laws in biology— only gadgets." This was particularly telling coming from the brilliant Nobel Prize winner, who had earlier presented the "central dogma" of molecular biology, implying that life might be explained just by the phrase "DNA makes RNA makes protein." Cited in A Varki, "Nothing in Glycobiology Makes Sense, Except in the Light of Evolution," *Cell* 126 (2006): 841–45.

17   HJ Jerison, "Animal Intelligence as Encephalization," *Philosophical Transactions of the Royal Society of London B: Biological Sciences* 308 (1985): 21–35; M Balter, "Mysteries of the Brain: Why Are Our Brains So Big?" *Science* 338 (2012): 33–34.

18   AH Taylor, DM Elliffe, GR Hunt, NJ Emery, NS Clayton, and RD Gray, "New Caledonian Crows Learn the Functional Properties of Novel Tool Types," *PLoS ONE* 6 (2011): e26887; Tim Birkhead, *Bird Sense: What It's Like to Be a Bird* (Walker & Company, 2012); John Marzluff and Tony Angell, *Gifts of the Crow: How Percep-*

tion, Emotion, and Thought Allow Smart Birds to Behave Like Humans (Free Press, 2012).

19   JS Reilly, EA Bates, and VA Marchman, "Narrative Discourse in Children with Early Focal Brain Injury," *Brain and Language* 61 (1998): 335–75; MA Ciliberto, D Limbrick, A Powers, JB Titus, R Munro, and MD Smyth, "Palliative Hemispherotomy in Children with Bilateral Seizure Onset," *Journal of Neurosurgery: Pediatrics* 9 (2012): 381–88.

20   Michael S. Gazzaniga, *Human: The Science Behind What Makes Us Unique* (Ecco, 2008).

21   Stephen R. Anderson, *Doctor Dolittle's Delusion: Animals and the Uniqueness of Human Language* (Yale University Press, 2004).

22   Philip Lieberman, *Eve Spoke: Human Language and Human Evolution* (W. W. Norton, 1998); P Lieberman, ES Crelin, and DH Klatt, "Phonetic Ability and Related Anatomy of the Newborn and Adult Human, Neanderthal Man, and the Chimpanzee," *American Anthropologist* 74 (2009): 287–307.

23   SE Fisher, CS Lai, and AP Monaco, "Deciphering the Genetic Basis of Speech and Language Disorders," *Annual Review of Neuroscience* 26 (2003): 57–80; SA White, SE Fisher, DH Geschwind, C Scharff, and TE Holy, "Singing Mice, Songbirds, and More: Models for FOXP2 Function and Dysfunction in Human Speech and Language," *Journal of Neuroscience* 26 (2006): 10376–79; W Enard, "Foxp2 and the Role of Cortico-Basal Ganglia Circuits in Speech and Language Evolution," *Current Opinion in Neurobiology* 21 (2011): 415–24.

24   Francine Patterson and Eugene Linden, *The Education of Koko* (Henry Holt, 1988); Sue Savage-Rumbaugh, Stuart G. Shanker, and Talbot J. Taylor, *Apes, Language, and the Human Mind* (Oxford University Press, 2001); Sally Boysen and Deborah Custance, *The Smartest Animals on the Planet: Extraordinary Tales of the Natural World's Cleverest Creatures* (Firefly Books, 2009).

25   Diana Reiss, *The Dolphin in the Mirror: Exploring Dolphin Minds and Saving Dolphin Lives* (Houghton Mifflin Harcourt, 2011).

26   Krish Sathian of Emory University pointed out to me that language understanding is clearly multisensory, with important visual contributions—witness the McGurk effect—and that language ac-

cesses distributed, multimodal brain representations (visual, emotional, motor, etc.) regardless of the route of entry.

27   MV Olson, EE Eichler, A Varki, RM Myers, JM Erwin, and EH McConkey, "Advocating Complete Sequencing of the Genome of the Common Chimpanzee, *Pan troglodytes*"; DE Reich, ES Lander, R Waterston, S Pääbo, M Ruvolo, and A Varki, "Sequencing the Chimpanzee Genome" (white papers submitted to the National Human Genome Research Institute, February 2002).

28   The Chimpanzee Genome Sequencing Consortium, "Initial Sequence of the Chimpanzee Genome and Comparison with the Human Genome," *Nature* 437 (2005): 69–87.

29   M O'Bleness, VB Searles, A Varki, P Gagneux, and JM Sikela, "Evolution of Genetic and Genomic Features Unique to the Human Lineage," *Nature Reviews: Genetics* 13, no. 12 (2012): 853–66.

30   KS Pollard, SR Salama, N Lambert, MA Lambot, S Coppens, JS Pedersen, S Katzman, B King, C Onodera, A Siepel, AD Kern, C Dehay, H Igel, M Ares Jr., P Vanderhaeghen, and D Haussler, "An RNA Gene Expressed during Cortical Development Evolved Rapidly in Humans," *Nature* 443, no. 7108 (2006): 167–72; HA Burbano, RE Green, T Maricic, C Lalueza-Fox, M de la Rasilla, A Rosas, J Kelso, KS Pollard, M Lachmann, and S Pääbo, "Analysis of Human Accelerated DNA Regions Using Archaic Hominin Genomes," *PLoS ONE* 7, no. 3 (2012): e32877.

31   HH Stedman, BW Kozyak, A Nelson, DM Thesier, LT Su, DW Low, CR Bridges, JB Shrager, N Minugh-Purvis, and MA Mitchell, "Myosin Gene Mutation Correlates with Anatomical Changes in the Human Lineage," *Nature* 428 (2004): 415–18.

32   GH Perry, BC Verrelli, and AC Stone, "Comparative Analyses Reveal a Complex History of Molecular Evolution for Human MYH16," *Molecular Biology and Evolution* 22 (2005): 379–82; MA McCollum, CC Sherwood, CJ Vinyard, CO Lovejoy, and F Schachat, "Of Muscle-Bound Crania and Human Brain Evolution: The Story Behind the MYH16 Headlines," *Journal of Human Evolution* 50 (2006): 232–36.

33   T Hayakawa, T Angata, AL Lewis, TS Mikkelsen, NM Varki, and A Varki, "A Human-Specific Gene in Microglia," *Science* 309 (2005): 1693; X Wang, N Mitra, P Cruz, L Deng, N Varki, T Angata,

ED Green, J Mullikin, T Hayakawa, and A Varki, "Evolution of SIGLEC-11 and SIGLEC-16 Genes in Hominins," *Molecular Biology and Evolution* 29 (2012): 2073–86.

34 Rudyard Kipling, *Just So Stories for Little Children* (Doubleday, Page & Co., 1902).

35 W Enard, P Khaitovich, J Klose, S Zollner, F Heissig, P Giavalisco, K Nieselt-Struwe, E Muchmore, A Varki, R Ravid, GM Doxiadis, RE Bontrop, and S Pääbo, "Intra- and Interspecific Variation in Primate Gene Expression Patterns," *Science* 296 (2002): 340–43; M Caceres, J Lachuer, MA Zapala, JC Redmond, L Kudo, DH Geschwind, DJ Lockhart, TM Preuss, and C Barlow, "Elevated Gene Expression Levels Distinguish Human from Non-Human Primate Brains," *Proceedings of the National Academy of Sciences of the United States of America* 100 (2003): 13030–35; G Konopka, T Friedrich, J Davis-Turak, K Winden, MC Oldham, F Gao, L Chen, GZ Wang, R Luo, TM Preuss, and DH Geschwind, "Human-Specific Transcriptional Networks in the Brain," *Neuron* 75 (2012): 601–17; TM Preuss, "Human Brain Evolution: From Gene Discovery to Phenotype Discovery," *Proceedings of the National Academy of Sciences of the United States of America* 109, suppl 1 (2012): 10709–16.

36 MC King and AC Wilson, "Evolution at Two Levels in Humans and Chimpanzees," *Science* 188 (1975): 107–16.

37 http://carta.anthropogeny.org/content/about-moca. The Matrix of Comparative Anthropogeny (MOCA) is a collection of comparative information regarding humans and our closest evolutionary cousins, the great apes. The emphasis is on uniquely human features.

38 Many writers have described and discussed this issue over the years. For examples, see these sources and the references therein: Mary Jane West-Eberhard, *Developmental Plasticity and Evolution* (Oxford University Press, 2003); Melvin Konner, *The Evolution of Childhood: Relationships, Emotion, Mind* (Belknap Press of Harvard University Press, 2010); Noël Cameron and Barry Bogin, eds., *Human Growth and Development,* 2nd ed. (Academic Press, 2012); HM Dunsworth, AG Warrener, T Deacon, PT Ellison, and H Pontzer, "Metabolic Hypothesis for Human Altriciality," *Pro-*

ceedings of the National Academy of Sciences of the United States of America 109 (2012): 15212–16.

39 Barry Bogin and Holly B. Smith, "Evolution of the Human Life Cycle," in *Human Biology: An Evolutionary and Biocultural Perspective*, edited by Sara Stinson, Barry Bogin, and Dennis O'Rourke (Wiley-Blackwell, 2012).

40 Even Neandertals, apparently, matured faster than humans. See TM Smith, P Tafforeau, DJ Reid, J Pouech, V Lazzari, JP Zermeno, D Guatelli-Steinberg, AJ Olejniczak, A Hoffman, J Radovcic, M Makaremi, M Toussaint, C Stringer, and JJ Hublin, "Dental Evidence for Ontogenetic Differences between Modern Humans and Neanderthals," *Proceedings of the National Academy of Sciences of the United States of America* 107 (2010): 20923–28.

41 Sarah Blaffer Hrdy, *Mothers and Others: The Evolutionary Origins of Mutual Understanding* (Belknap Press of Harvard University Press, 2009).

42 Stephen Jay Gould, *Wonderful Life: The Burgess Shale and the Nature of History* (W. W. Norton, 1990).

43 F Hailer, VE Kutschera, BM Hallstrom, D Klassert, SR Fain, JA Leonard, U Arnason, and A Janke, "Nuclear Genomic Sequences Reveal That Polar Bears Are an Old and Distinct Bear Lineage," *Science* 336, no. 6079 (2012): 344–47. With the melting of Arctic ice due to global warming, the ranges of brown bears and polar bears are now overlapping, so an unnatural experiment in rehybridization is under way.

44 C Vila, P Savolainen, JE Maldonado, IR Amorim, JE Rice, RL Honeycutt, KA Crandall, J Lundeberg, and RK Wayne, "Multiple and Ancient Origins of the Domestic Dog," *Science* 276 (1997): 1687–89; HG Parker, AL Shearin, and EA Ostrander, "Man's Best Friend Becomes Biology's Best in Show: Genome Analyses in the Domestic Dog," *Annual Review of Genetics* 44 (2010): 309–36; RK Wayne and BM vonHoldt, "Evolutionary Genomics of Dog Domestication," *Mammalian Genome* 23 (2012): 3–18.

45 Y Gilad, CD Bustamante, D Lancet, and S Pääbo, "Natural Selection on the Olfactory Receptor Gene Family in Humans and Chimpanzees," *American Journal of Human Genetics* 73 (2003): 489–501; Y Gilad, O Man, S Pääbo, and D Lancet, "Human Specific

Loss of Olfactory Receptor Genes," *Proceedings of the National Academy of Sciences of the United States of America* 100 (2003): 3324–27.

46   Wenda R. Trevathan, *Human Birth: An Evolutionary Perspective* (Aldine Transaction, 2011).

47   Naomi Oreskes, *The Rejection of Continental Drift: Theory and Method in American Earth Science* (Oxford University Press, 1999).

Chapter 3. There Are No Free Lunches or Free Smarts

1   Chip Walter, *Thumbs, Toes, and Tears: And Other Traits That Make Us Human* (Walker & Company, 2008).

2   A Varki, DH Geschwind, and EE Eichler, "Explaining Human Uniqueness: Genome Interactions with Environment, Behaviour, and Culture," *Nature Reviews Genetics* 9 (2008): 749–63. The second part of this review contains many ideas relevant to this book.

3   SG Brickley, EA Dawes, MJ Keating, and S Grant, "Synchronizing Retinal Activity in Both Eyes Disrupts Binocular Map Development in the Optic Tectum," *Journal of Neuroscience* 18 (1998): 1491–504.

4   Daniel Kahneman, *Thinking, Fast and Slow* (Farrar, Straus and Giroux, 2011).

5   Joseph E. LeDoux, *The Emotional Brain: The Mysterious Underpinnings of Emotional Life* (Simon and Schuster, 1998); Paul Ekman, *Emotions Inside Out: 130 Years after Darwin's* The Expression of the Emotions in Man and Animals, *Annals of the New York Academy of Sciences* 1000 (2003); Daniel Goleman, *Emotional Intelligence: Why It Can Matter More Than IQ*, 10th anniversary ed. (Bantam Books, 2005).

6   S. Boyd Eaton, Marjorie Shostak, and Melvin Konner, *The Paleolithic Prescription* (Harper & Row, 1988); Peter Gluckman and Mark Hanson, *Mismatch: The Lifestyle Diseases Timebomb* (Oxford University Press, 2008).

7   DA Treffert, "The Savant Syndrome: An Extraordinary Condition.

A Synopsis: Past, Present, Future," *Philosophical Transactions of the Royal Society B: Biological Sciences* 364 (2009): 1351–57.

8   Francesca Happé and Uta Frith, eds., *Autism and Talent* (Oxford University Press, 2010); Michael Fitzgerald, *Autism and Creativity: Is There a Link between Autism in Men and Exceptional Ability?* (Brunner-Routledge, 2004).

9   WA Horwitz, C Kestenbaum, E Person, and L Jarvik, "Identical Twins—'Idiot Savants'—Calendar Calculators," *American Journal of Psychiatry* 121 (1965): 1075–79.

10   JR Hughes, "The Savant Syndrome and Its Possible Relationship to Epilepsy," *Advances in Experimental Medicine and Biology* 724 (2012): 332–43.

11   BL Miller, J Cummings, F Mishkin, K Boone, F Prince, M Ponton, and C Cotman, "Emergence of Artistic Talent in Frontotemporal Dementia," *Neurology* 51, no. 4 (1998): 978–82.

12   ES Parker, L Cahill, and JL McGaugh, "A Case of Unusual Autobiographical Remembering," *Neurocase* 12 (2006): 35–49; BA Ally, EP Hussey, and MJ Donahue, "A Case of Hyperthymesia: Rethinking the Role of the Amygdala in Autobiographical Memory," *Neurocase* (2012): Epub ahead of print.

13   Joaquin Fuster, *The Prefrontal Cortex*, 4th ed. (Academic Press, 2008).

14   Temple Grandin and Catherine Johnson, *Animals in Translation: Using the Mysteries of Autism to Decode Animal Behavior* (Scribner, 2005).

15   T Grandin, "How Does Visual Thinking Work in the Mind of a Person with Autism? A Personal Account," *Philosophical Transactions of the Royal Society B: Biological Sciences* 364 (2009): 1437–42.

Chapter 4. Many Levels of Awareness

1   GG Gallup Jr., "Absence of Self-Recognition in a Monkey (*Macaca fascicularis*) following Prolonged Exposure to a Mirror," *Developmental Psychobiology* 10 (1977): 281–84.

2   Sue Taylor Parker, Robert W. Mitchell, and Maria L. Boccia, *Self-*

Awareness in Animals and Humans: Developmental Perspectives (Cambridge University Press, 1994).

3   Derek Denton, *The Primordial Emotions: The Dawning of Consciousness* (Oxford University Press, 2005).

4   A Kitchen, D Denton, and L Brent, "Self-Recognition and Abstraction Abilities in the Common Chimpanzee Studied with Distorting Mirrors," *Proceedings of the National Academy of Sciences of the United States of America* 93 (1996): 7405–08.

5   Julian Paul Keenan, Gordon G. Gallup Jr., and Dean Falk, *The Face in the Mirror: The Search for the Origins of Consciousness* (Ecco Press, 2003).

6   Daniel J. Povinelli, Timothy J. Eddy, R. Peter Hobson, and Michael Tomasello, *What Young Chimpanzees Know about Seeing* (University of Chicago Press, 1996).

7   DJ Shillito, GG Gallup Jr., and BB Beck, "Factors Affecting Mirror Behaviour in Western Lowland Gorillas, *Gorilla gorilla*," *Animal Behaviour* 57 (1999): 999–1004.

8   The term *intentionality* was first introduced by the philosopher Franz Brentano (1838–1917) and later modified into multiple levels by the American philosopher Daniel C. Dennett (*The Intentional Stance*, MIT Press, 1987). It has also been applied by Robin Dunbar to the social brain hypothesis.

9   Josep Call and Michael Tomasello, "Does the Chimpanzee Have a Theory of Mind? 30 Years Later," *Trends in Cognitive Sciences* 12 (2008): 187–92. Based on this paper, the following definitions can be considered: self-awareness is "awareness of oneself as an intentional agent"; a rudimentary theory of mind is "understanding others as intentional agents in terms of a perception-goal psychology"; a full theory of mind is "understanding others as intentional agents in a fully humanlike belief-desire psychology and appreciating that others have mental representations of the world that drive their actions, even when those do not correspond to reality."

10 Perhaps we can use a more precise term, like *awareness of oneself*. In other words, a self-aware individual is aware that he or she is an individual person. In this terminology one can then replace *theory of mind* with *awareness of another self*. The next stage would then be *awareness of multiple other selves*. Regardless of

whether you prefer to stay with the old terms (as we have done in this book) or use some of these new ideas, two things need to be made clear. First, there is no clear-cut line between these different stages of awareness; they are just arbitrary stages on a continuum. Second, since self-awareness seems to have appeared during animal evolution on multiple independent occasions, we should not assume that the neural mechanisms involved are the same in every case. It might well be that a dolphin's awareness of self uses a different neural mechanism from a chimpanzee's. Regardless of which terms you may choose, it appears that becoming fully aware of the minds of others and their personhood is something that only human beings have achieved.

11   D Premack and G Woodruff, "Does the Chimpanzee Have a Theory of Mind?" *Behavioral and Brain Sciences* 4 (1978): 515–26.

12   Arguments over meanings of terms tend to get scientists very excited and easily upset. So it is wise for us to steer clear of this fray. Besides, we are not published experts on this topic. But for aficionados only, here is a possible solution, which retains parts of the original terms, to consider for the future. Perhaps theory of mind includes (as arbitrary stages on a continuum):

• Theory of one's own mind (TOOM; self-awareness; first-order intentionality)
• Theory of another mind (TAM; awareness of another individual as an intentional agent; second-order intentionality)
• Theory of other minds (TOM; full awareness of the self-awareness of others; third-order intentionality)

13   Emma Townshend, *Darwin's Dogs: How Darwin's Pets Helped Form a World-Changing Theory of Evolution* (Frances Lincoln, 2009).

14   John Bradshaw, *Dog Sense: How the New Science of Dog Behavior Can Make You a Better Friend to Your Pet* (Basic Books, 2011).

15   B Hare and M Tomasello, "Human-Like Social Skills in Dogs?" *Trends in Cognitive Science* 9 (2005): 439–44.

16   H Kobayashi and S Kohshima, "Unique Morphology of the Human Eye," *Nature* 387 (1997): 767–68.

17   M Tomasello, B Hare, H Lehmann, and J Call, "Reliance on Head versus Eyes in the Gaze Following of Great Apes and Human

Infants: The Cooperative Eye Hypothesis," *Journal of Human Evolution* 52, no. 3 (2007): 314–20.

18 Jeffrey Moussaieff Masson, *The Dog Who Couldn't Stop Loving: How Dogs Have Captured Our Hearts for Thousands of Years* (Harper Paperbacks, 2011).

19 B Hare and M Tomasello, "Human-Like Social Skills in Dogs?" *Trends in Cognitive Science* 9 (2005): 439–44.

20 L Trut, I Oskina, and A Kharlamova, "Animal Evolution during Domestication: The Domesticated Fox as a Model," *BioEssays* 31 (2009): 349–60.

21 M Germonpré, MV Sablin, RE Stevens, REM Hedges, M Hofreiter, M Stiller, and VR Després, "Fossil Dogs and Wolves from Palaeolithic Sites in Belgium, the Ukraine, and Russia: Osteometry, Ancient DNA, and Stable Isotopes," *Journal of Archaeological Science* 36 (2009): 473–90; ND Ovodov, SJ Crockford, YV Kuzmin, TF Higham, GW Hodgins, and J van der Plicht, "A 33,000-Year-Old Incipient Dog from the Altai Mountains of Siberia: Evidence of the Earliest Domestication Disrupted by the Last Glacial Maximum," *PLoS ONE* 6 (2011): e22821.

22 http://carta.anthropogeny.org/content/about-moca. The Matrix of Comparative Anthropogeny (MOCA) is a collection of comparative information regarding humans and our closest evolutionary cousins, the great apes. The emphasis is on uniquely human features.

23 Donald Brown, *Human Universals* (McGraw-Hill, 1991). See also Peter M. Kappeler and Joan B. Silk, eds., *Mind the Gap: Tracing the Origins of Human Universals* (Springer, 2010).

24 Matthew M. Hurley and Daniel C. Dennett, *Inside Jokes: Using Humor to Reserve-Engineer the Mind* (MIT Press, 2001).

25 "Mental time travel" refers to the conscious experience of remembering the personal past and imagining the personal future. See L Nyberg, AS Kim, R Habib, B Levine, and E Tulving, "Consciousness of Subjective Time in the Brain," *Proceedings of the National Academy of Sciences of the United States of America* 107 (2010): 22356–59.

26 J Quoidbach, DT Gilbert, and TD Wilson, "The End of History Illusion," *Science* 339, no. 6115 (2013): 96–98.

27 Richard W. Wrangham, *Catching Fire: How Cooking Made Us Human* (New York: Basic Books, 2009).

28   K Hawkes, JF O'Connell, NG Jones, H Alvarez, and EL Charnov, "Grandmothering, Menopause, and the Evolution of Human Life Histories," *Proceedings of the National Academy of Sciences of the United States of America* 95 (1998): 1336–39; K Hawkes, "Colloquium Paper: How Grandmother Effects Plus Individual Variation in Frailty Shape Fertility and Mortality: Guidance From Human-Chimpanzee Comparisons," *Proceedings of the National Academy of Sciences of the United States of America* 107, suppl. 2 (2010): 8977–84; PS Kim, JE Coxworth, and K Hawkes, "Increased Longevity Evolves from Grandmothering," *Proceedings of the Royal Society B: Biological Sciences* 279 (2012): 4880–84.

29   K Hill and AM Hurtado, "Social Science: Human Reproductive Assistance," *Nature* 483 (2012): 160–61.

30   EA Foster, DW Franks, S Mazzi, SK Darden, KC Balcomb, JK Ford, and DP Croft, "Adaptive Prolonged Postreproductive Life Span in Killer Whales," *Science* 337 (2012): 1313.

31   JD Maser and GG Gallup Jr., "Theism as a By-Product of Natural Selection," *The Journal of Religion* 70 (1990): 515–32; Pascal Boyer, *Religion Explained: The Evolutionary Origins of Religious Thought* (Basic Books, 2001); David Sloan Wilson, *Darwin's Cathedral: Evolution, Religion, and the Nature of Society* (University of Chicago Press, 2002); Daniel C. Dennett, *Breaking the Spell: Religion as a Natural Phenomenon* (Viking, 2006); P Boyer, "Being Human: Religion: Bound to Believe?" *Nature* 455 (2008): 1038–33; A Norenzayan and AF Shariff, "The Origin and Evolution of Religious Prosociality," *Science* 322 (2008): 58–62; Nicholas Wade, *The Faith Instinct: How Religion Evolved and Why It Endures* (Penguin Press, 2009); Bruce M. Hood, *Supersense: Why We Believe in the Unbelievable* (HarperOne, 2009); Jesse Bering, *The Belief Instinct: The Psychology of Souls, Destiny, and the Meaning of Life* (W. W. Norton, 2011); Robert N. McCauley, *Why Religion Is Natural and Science Is Not* (Oxford University Press, 2011).

32   Sam Harris, *The End of Faith: Religion, Terror, and the Future of Reason* (W. W. Norton, 2005); Daniel C. Dennett, *Breaking the Spell: Religion as a Natural Phenomenon* (Viking, 2006); Richard Dawkins, *The God Delusion* (Houghton Mifflin, 2006); Christo-

pher Hitchens, *God Is Not Great: How Religion Poisons Everything* (Twelve, 2009).

33   Helen Fisher, *Why We Love: The Nature and Chemistry of Romantic Love* (Holt Paperbacks, 2005).

34   Matt Ridley, *The Rational Optimist: How Prosperity Evolves* (Harper, 2010).

35   AN Meltzoff, PK Kuhl, J Movellan, and TJ Sejnowski, "Foundations for a New Science of Learning," *Science* 325 (2009): 284–88.

36   Martin Nowak and Roger Highfield, *SuperCooperators: Altruism, Evolution, and Why We Need Each Other to Succeed* (Free Press, 2011).

37   Edward O. Wilson, *The Social Conquest of Earth* (Liveright, 2012).

38   Derek Denton, *The Primordial Emotions: The Dawning of Consciousness* (Oxford University Press, 2005).

39   Francis Crick, *Astonishing Hypothesis: The Scientific Search for the Soul* (Scribner, 1995).

40   Gerald Edelman, *The Remembered Present: A Biological Theory of Consciousness* (Basic Books, 1990); Gerald M. Edelman, *Wider Than the Sky: The Phenomenal Gift of Consciousness* (Yale University Press, 2005).

41   Merlin Donald, *A Mind So Rare: The Evolution of Human Consciousness* (W. W. Norton, 2001).

42   Daniel C. Dennett, *Consciousness Explained* (Penguin Books, 2007).

43   Nicholas Humphrey, *Soul Dust: The Magic of Consciousness* (Princeton University Press, 2011).

44   Antonio R. Damasio, *Self Comes to Mind: Constructing the Conscious Brain* (Pantheon Books, 2010).

45   Christof Koch, *Consciousness: Confessions of a Romantic Reductionist* (MIT Press, 2012).

46   Thomas Metzinger, *The Ego Tunnel: The Science of the Mind and the Myth of the Self* (Basic Books, 2010).

47   Giacomo Rizzolatti and Corrado Sinigaglia, *Mirrors in the Brain: How Our Minds Share Actions and Emotions* (Oxford University Press, 2008); Marco Iacoboni, *Mirroring People: The Science of Empathy and How We Connect with Others* (Farrar, Straus and Giroux, 2008); Frans B. M. de Waal and Pier Francesco Ferrari,

eds., *The Primate Mind: Built to Connect with Other Minds* (Harvard University Press, 2012).

48   MJ Banissy and J Ward, "Mirror-Touch Synesthesia Is Linked with Empathy," *Nature Neuroscience* 10 (2007): 815–16.

49   VS Ramachandran, *The Tell-Tale Brain: A Neuroscientist's Quest for What Makes Us Human* (W. W. Norton, 2011).

50   Michael A. Arbib, *How the Brain Got Language: The Mirror System Hypothesis* (Oxford University Press, 2012).

51   Y Li, H Lu, PL Cheng, S Ge, H Xu, SH Shi, and Y Dan, "Clonally Related Visual Cortical Neurons Show Similar Stimulus Feature Selectivity," *Nature* 486 (2012):118–121.

52   Uta Frith, *Autism: Explaining the Enigma* (Blackwell, 2003); Laura Ellen Schreibman, *The Science and Fiction of Autism* (Harvard University Press, 2005); Daniel Tammet, *Born on a Blue Day: Inside the Extraordinary Mind of an Autistic Savant* (Free Press, 2007); Chloe Silverman, *Understanding Autism: Parents, Doctors, and the History of a Disorder* (Princeton University Press, 2012).

53   S Baron-Cohen, AM Leslie, and U Frith, "Does the Autistic Child Have a 'Theory of Mind'?" *Cognition* 21 (1985): 37–46.

54   This is a summary of the proposed criteria as of late 2012—a final version is due out in 2013.

55   S Baron-Cohen, M O'Riordan, V Stone, R Jones, and K Plaisted, "Recognition of Faux Pas by Normally Developing Children and Children with Asperger Syndrome or High-Functioning Autism," *Journal of Autism and Developmental Disorders* 29 (1999): 407–18.

56   F Happé and U Frith, "The Beautiful Otherness of the Autistic Mind," *Philosophical Transactions of the Royal Society B: Biological Sciences* 364 (2009): 1346–50.

57   L Mottron, "Changing Perceptions: The Power of Autism," *Nature* 479 (2011): 33–35.

58   http://specialistpeople.com/?id=159; http://usa.specialisterne.com/about-specialisterne/specialist-people-foundation/.

59   X Zhao, A Leotta, V Kustanovich, C Lajonchere, DH Geschwind, K Law, P Law, S Qiu, C Lord, J Sebat, K Ye, and M Wigler, "A Unified Genetic Theory for Sporadic and Inherited Autism," *Proceedings of the National Academy of Sciences of the United States*

of America 104 (2007): 12831–36; C Lord, "Unweaving the Autism Spectrum," *Cell* 147 (2011): 24–25; MW State and N Sestan, "Neuroscience: The Emerging Biology of Autism Spectrum Disorders," *Science* 337 (2012): 1301–03.

60 A Kong, ML Frigge, G Masson, S Besenbacher, P Sulem, G Magnusson, SA Gudjonsson, A Sigurdsson, A Jonasdottir, A Jonasdottir, WS Wong, G Sigurdsson, GB Walters, S Steinberg, H Helgason, G Thorleifsson, DF Gudbjartsson, A Helgason, OT Magnusson, U Thorsteinsdottir, and K Stefansson, "Rate of De Novo Mutations and the Importance of Father's Age to Disease Risk," *Nature* 488 (2012): 471–75.

61 MT Roelfsema, RA Hoekstra, C Allison, S Wheelwright, C Brayne, FE Matthews, and S Baron-Cohen, "Are Autism Spectrum Conditions More Prevalent in an Information-Technology Region? A School-Based Study of Three Regions in the Netherlands," *Journal of Autism and Developmental Disorders* 42 (2012): 734 39.

62 Simon Baron-Cohen, *The Essential Difference: Men, Women and the Extreme Male Brain* (Penguin Books, 2007).

63 JT Morgan, G Chana, I Abramson, K Semendeferi, E Courchesne, and IP Everall, "Abnormal Microglial-Neuronal Spatial Organization in the Dorsolateral Prefrontal Cortex in Autism," *Brain Research* 1456 (2012): 72–81.

64 C Chevallier, G Kohls, V Troiani, ES Brodkin, and RT Schultz, "The Social Motivation Theory of Autism," *Trends in Cognitive Sciences* 16 (2012): 231–39.

65 B Crespi, P Stead, and M Elliot, "Evolution in Health and Medicine Sackler Colloquium: Comparative Genomics of Autism and Schizophrenia," *Proceedings of the National Academy of Sciences of the United States of America* 107, suppl. 1 (2010): 1736–41.

Chapter 5. The Wall

1 H Li and R Durbin, "Inference of Human Population History from Individual Whole-Genome Sequences," *Nature* 475 (2011): 493–96.

2 Sarah Blaffer Hrdy, *Mothers and Others: The Evolutionary Origins*

of Mutual Understanding (Belknap Press of Harvard University Press, 2009).

3   Christopher Boehm, *Moral Origins: The Evolution of Virtue, Altruism, and Shame* (Basic Books, 2012).

4   Patricia S. Churchland, *Braintrust: What Neuroscience Tells Us about Morality* (Princeton University Press, 2011).

5   One might argue that acceptance, not denial, is at least a plausible alternative. But one would be arguing from the point of view of a modern human, not from that of the first rare individuals who achieved the transition.

6   Robert Trivers, *The Folly of Fools: The Logic of Deceit and Self-Deception in Human Life* (Basic Books, 2011). Ian Leslie also argues that lying is central to who we are as humans in *Born Liars: Why We Can't Live Without Deceit* (Quercus, 2011).

7   Robert Trivers, *Natural Selection and Social Theory: Selected Papers of Robert L. Trivers* (Oxford University Press, 2002).

8   K. V. and Lali Krishnan pointed me to these concepts.

9   A Rosenblatt, J Greenberg, S Solomon, T Pyszczynski, and D Lyon, "Evidence for Terror Management Theory: I. The Effects of Mortality Salience on Reactions to Those Who Violate or Uphold Cultural Values," *Journal of Personality and Social Psychology* 57 (1989): 681–90; S Solomon, J Greenberg, and T Pyszczynski, "A Terror Management Theory of Social Behavior: The Psychological Functions of Self-Esteem and Cultural Worldviews," *Advances in Experimental Social Psychology* 24 (1991): 93–159; J Greenberg, S Solomon, and T Pyszczynski, "Terror Management Theory of Self-Esteem and Cultural Worldviews: Empirical Assessments and Conceptual Refinements," *Advances in Experimental Social Psychology* 29 (1997): 61–139; T Pyszczynski, J Greenberg, and S Solomon, "A Dual-Process Model of Defense against Conscious and Unconscious Death-Related Thoughts: An Extension of Terror Management Theory," *Psychological Review* 106 (1999): 835–45.

10   VS Ramachandran, "The Evolutionary Biology of Self-Deception, Laughter, Dreaming, and Depression: Some Clues from Anosognosia," *Medical Hypotheses* 47 (1996): 347–62.

11   Richard P. Feynman, *What Do You Care What Other People Think? Further Adventures of a Curious Character* (W. W. Norton, 2001).

12 Robert Langs, *Death Anxiety and Clinical Practice* (Karnac Books, 1997); *Psychotherapy and Science* (Sage, 1999); *Freud on a Precipice: How Freud's Fate Pushed Psychoanalysis over the Edge* (Jason Aronson, 2010).

13 Stephen Cave, *Immortality: The Quest to Live Forever and How It Drives Civilization* (Crown, 2012).

14 Nicholas Humphrey, *Soul Dust: The Magic of Consciousness* (Princeton University Press, 2011).

15 T Sharot, AM Riccardi, CM Raio, and EA Phelps, "Neural Mechanisms Mediating Optimism Bias," *Nature* 450, no. 7166 (2007): 102–5; T Sharot, CW Korn, and RJ Dolan, "How Unrealistic Optimism Is Maintained in the Face of Reality," *Nature Neuroscience* 14 (2011): 1475–79; T Sharot, R Kanai, D Marston, CW Korn, G Rees, and RJ Dolan, "Selectively Altering Belief Formation in the Human Brain," *Proceedings of the National Academy of Sciences of the United States of America* 109 (2012): 17058–62.

16 DJ Miller, T Duka, CD Stimpson, SJ Schapiro, WB Baze, MJ McArthur, AJ Fobbs, AM Sousa, N Sestan, DE Wildman, L Lipovich, CW Kuzawa, PR Hof, and CC Sherwood, "Prolonged Myelination in Human Neocortical Evolution," *Proceedings of the National Academy of Sciences of the United States of America* 109 (2012): 16480–85.

17 Melvin Konner, *The Evolution of Childhood: Relationships, Emotion, Mind* (Belknap Press of Harvard University Press, 2010); Alison Gopnik, *The Philosophical Baby: What Children's Minds Tell Us about Truth, Love, and the Meaning of Life* (Farrar, Straus and Giroux, 2009).

18 MW Speece and SB Brent, "Children's Understanding of Death: A Review of Three Components of a Death Concept," *Child Development* 55 (1984): 1671–86; SP Nguyen and SA Gelman, "Four- and Six-Year Olds' Concept of Death," *British Journal of Developmental Psychology* 20 (2002): 495–513; HC Barrett and T Behne, "Children's Understanding of Death as the Cessation of Agency: A Test Using Sleep Versus Death," *Cognition* 96 (2005): 93–108.

19 I Dumontheil, IA Apperly, and SJ Blakemore, "Online Usage of Theory of Mind Continues to Develop in Late Adolescence," *Developmental Science* 13 (2010): 331–38; SJ Blakemore and TW

Robbins, "Decision-Making in the Adolescent Brain," *Nature Neuroscience* 15 (2012): 1184–91.

20   Barry Bogin and Holly B. Smith, "Evolution of the Human Life Cycle," in *Human Biology: An Evolutionary and Biocultural Perspective*, edited by Sara Stinson, Barry Bogin, and Dennis O'Rourke (Wiley-Blackwell, 2012).

21   SJ Gould and N Eldredge, "Punctuated Equilibrium Comes of Age," *Nature* 366 (1993): 223–27.

22   Sewall Wright, "Surfaces of Selective Value Revisited," *The American Naturalist* 131 (1988): 115–23.

Chapter 6. Breaking through the Wall

1   C Crockford, RM Wittig, R Mundry, and K Zuberbuhler, "Wild Chimpanzees Inform Ignorant Group Members of Danger," *Current Biology* 22 (2012): 142–46; RM Seyfarth and DL Cheney, "Animal Cognition: Chimpanzee Alarm Calls Depend on What Others Know," *Current Biology* 22 (2012): R51–52.

2   JM Dally, NJ Emery, and NS Clayton, "Food-Caching Western Scrub-Jays Keep Track of Who Was Watching When," *Science* 312 (2006): 1662–65; NJ Emery and NS Clayton, "Comparative Social Cognition," *Annual Review of Psychology* 60 (2009): 87–113.

3   M Watve, J Thakar, A Kale, S Puntambekar, I Shaikh, K Vaze, M Jog, and S Paranjape, "Bee-Eaters (*Merops orientalis*) Respond to What a Predator Can See," *Animal Cognition* 5 (2002): 253–59.

4   LA Bates, JH Poole, and RW Byrne, "Elephant Cognition," *Current Biology* 18 (2008): R544–46.

5   Cynthia J. Moss, Harvey Croze, and Phyllis C. Lee, *The Amboseli Elephants: A Long-Term Perspective on a Long-Lived Mammal* (University of Chicago Press, 2011).

6   R Slotow, G van Dyk, J Poole, B Page, and A Klocke, "Older Bull Elephants Control Young Males," *Nature* 408 (2000): 425–26; GA Bradshaw, AN Schore, JL Brown, JH Poole, and CJ Moss, "Elephant Breakdown," *Nature* 433 (2005): 807.

7   Cynthia J. Moss, *Elephant Memories: Thirteen Years in the Life of an Elephant Family* (University of Chicago Press, 2000); LA Bates,

KN Sayialel, NW Njiraini, CJ Moss, JH Poole, and RW Byrne, "Elephants Classify Human Ethnic Groups by Odor and Garment Color," *Current Biology* 17 (2007): 1938–42; Lawrence Anthony and Graham Spence, *The Elephant Whisperer: My Life with the Herd in the African Wild,* reprint ed. (St. Martin's Griffin, 2012).

8   JR Anderson, A Gillies, and LC Lock, "Pan Thanatology," *Current Biology* 20 (2010): R349–51; FA Stewart, AK Piel, and RC O'Malley, "Responses of Chimpanzees to a Recently Dead Community Member at Gombe National Park, Tanzania," *American Journal of Primatology* 74 (2012): 1–7.

9   D Biro, T Humle, K Koops, C Sousa, M Hayashi, and T Matsuzawa, "Chimpanzee Mothers at Bossou, Guinea, Carry the Mummified Remains of Their Dead Infants," *Current Biology* 20 (2010): R351–52.

10   Sharon Brower wrote me that "when Danny died, one of our dogs sat at the window for months waiting for him to come, and wailed at the top of his lungs (a sound he had never made before) every time he heard Danny's voice on the answering machine. It was so heartbreaking that I had to remove it. Until the day the dog died, I don't think that he ever got over losing Danny."

11   Julian Barnes, *Nothing to Be Frightened Of* (Knopf, 2009).

12   Of course, as an atheist friend pointed out to me, "nonconformist thinkers do not need the validation of others to reinforce their own beliefs. Perhaps they simply don't like conventions filled with preachy, organized thought that is reminiscent of religion. As an atheist, I never once felt the need or desire to go to a convention with other people who shared my beliefs. What would be the point of doing that?"

13   Simcha Paul Raphael, *Jewish Views of the Afterlife* (Rowman & Littlefield, 2009).

14   Takeru Akazawa, Kenichi Aoki, and Ofer Bar-Yosef, *Neandertals and Modern Humans in Western Asia* (Springer, 1998); CB Stringer, R Grun, HP Schwarcz, and P Goldberg, "ESR Dates for the Hominid Burial Site of Es Skhul in Israel," *Nature* 338 (1989): 756–58; M Vanhaereny, F d'Errico, C Stringer, SL James, JA Todd, and HK Mienis, "Middle Paleolithic Shell Beads in Israel and Algeria," *Science* 312 (2006): 1785–88; Paul Mellars, *The Ne-*

anderthal Legacy: An Archaeological Perspective from Western Europe (Princeton University Press, 1996); Paul Pettitt, *The Palaeolithic Origins of Human Burial* (Routledge, 2010).

15 M Balter, "Archaeology: Did Neandertals Truly Bury Their Dead?" *Science* 337 (2012): 1443–44.

16 C Finlayson, K Brown, R Blasco, J Rosell, JJ Negro, GR Bortolotti, G Finlayson, A Sanchez Marco, F Giles Pacheco, J Rodriguez Vidal, JS Carrion, DA Fa, and JM Rodriguez Llanes, "Birds of a Feather: Neanderthal Exploitation of Raptors and Corvids," *PLoS ONE* 7 (2012): e45927.

17 Steven Mithen, *The Singing Neanderthals: The Origins of Music, Language, Mind, and Body* (Harvard University Press, 2007); Thomas Wynn and Frederick Coolidge, *How to Think Like a Neandertal* (Oxford University Press, 2011).

18 JN Booth, SA Koren, and MA Persinger, "Increased Feelings of the Sensed Presence and Increased Geomagnetic Activity at the Time of the Experience during Exposures to Transcerebral Weak Complex Magnetic Fields," *International Journal of Neuroscience* 115 (2005): 1053–79.

19 Communicated to me by Randolph Nesse, an expert in psychiatry and evolution.

20 Francis S. Collins, *The Language of God: A Scientist Presents Evidence for Belief* (Free Press, 2007).

21 Francisco Ayala, *Darwin's Gift to Science and Religion* (Joseph Henry Press, 2007).

22 Kenneth R. Miller, *Finding Darwin's God: A Scientist's Search for Common Ground between God and Evolution*, reprint ed. (HarperPerennial, 2007).

23 Michael Ruse, *Science and Spirituality: Making Room for Faith in the Age of Science* (Cambridge University Press, 2010).

24 Louis W. Perry, *Thank Evolution for God: Nature and God's Creations and Designs* (Xlibris, 2012).

25 Robert N. McCauley, *Why Religion Is Natural and Science Is Not* (Oxford University Press, 2011).

26 Paul Bloom, *Descartes' Baby: How the Science of Child Development Explains What Makes Us Human* (Basic Books, 2004).

27 JM Bering and BD Parker, "Children's Attributions of Intentions to

an Invisible Agent," *Developmental Psychology* 42, no. 2 (2006): 253–62.

28 JM Lowenstein, "Twelve Wise Men at the Vatican," *Nature* 299, no. 5882 (1982): 395.

29 "Fate of Mountain Glaciers in the Anthropocene: A Report by the Working Group Commissioned by the Pontifical Academy of Sciences" (L. Bengtsson, P. J. Crutzen, and V. Ramanathan, co-chairs), April 2011.

30 Yet the Catholic Church does not support using condoms to prevent the spread of AIDS, so they have limits when it comes to accepting science.

31 O Berton, CG Hahn, and ME Thase, "Are We Getting Closer to Valid Translational Models for Major Depression?" *Science* 338 (2012): 75–79; Samuel H. Barondes, *Mood Genes: Hunting for Origins of Mania and Depression* (W. H. Freeman, 1998).

32 DA Haaga and AT Beck, "Perspectives on Depressive Realism: Implications for Cognitive Theory of Depression," *Behaviour Research and Therapy* 33 (1995): 41–48; R Pacini, F Muir, and S Epstein, "Depressive Realism from the Perspective of Cognitive-Experiential Self-Theory," *Journal of Personality and Social Psychology* 74 (1998): 1056–68; MT Moore and DM Fresco, "Depressive Realism: A Meta-Analytic Review," *Clinical Psychology Review* 32 (2012): 496–509.

33 R Machado-Vieira, G Salvadore, N Diazgranados, and CA Zarate Jr., "Ketamine and the Next Generation of Antidepressants with a Rapid Onset of Action," *Pharmacology and Therapeutics* 123 (2009): 143–50.

34 JE Kim, SR Dager, and IK Lyoo, "The Role of the Amygdala in the Pathophysiology of Panic Disorder: Evidence from Neuroimaging Studies," *Biology of Mood and Anxiety Disorders* 2, no. 1 (2012): 20.

35 DM Stoff and J Mann, "Suicide Research," *Annals of the New York Academy of Sciences* 836 (1997): 1–11; Kay Redfield Jamison, *Night Falls Fast: Understanding Suicide* (Knopf, 1999); AH MacKenzie, "What Makes Us Human," *The American Biology Teacher* 69 (2007): 522; A Preti, "Suicide among Animals: A Review of Evidence," *Psychological Reports* 101 (2007): 831–48.

36   Diana Reiss, *The Dolphin in the Mirror: Exploring Dolphin Minds and Saving Dolphin Lives* (Houghton Mifflin Harcourt, 2011).

37   T Sharot, AM Riccardi, CM Raio, and EA Phelps, "Neural Mechanisms Mediating Optimism Bias," *Nature* 450, no. 7166 (2007): 102–5; T Sharot, CW Korn, and RJ Dolan, "How Unrealistic Optimism Is Maintained in the Face of Reality," *Nature Neuroscience* 14 (2011): 1475–79; T Sharot, R Kanai, D Marston, CW Korn, G Rees, and RJ Dolan, "Selectively Altering Belief Formation in the Human Brain," *Proceedings of the National Academy of Sciences of the United States of America* 109 (2012): 17058–62.

38   Daniel Gilbert, *Stumbling on Happiness* (Vintage, 2007).

39   Dan Ariely, *Predictably Irrational: The Hidden Forces That Shape Our Decisions* (Harper, 2008).

40   DD Johnson and JH Fowler, "The Evolution of Overconfidence," *Nature* 477 (2011): 317–20.

41   Richard S. Tedlow, *Denial: Why Business Leaders Fail to Look Facts in the Face—and What to Do about It* (Portfolio, 2010); Margaret Heffernan, *Willful Blindness: Why We Ignore the Obvious at Our Peril* (Walker, 2011).

42   A Norenzayan, WM Gervais, and KH Trzesniewski, "Mentalizing Deficits Constrain Belief in a Personal God," *PLoS ONE* 7 (2012): e36880.

43   Jesse Bering, *The Belief Instinct: The Psychology of Souls, Destiny, and the Meaning of Life* (W. W. Norton, 2011).

44   U Frith and C Frith, "Reputation Management: In Autism, Generosity Is Its Own Reward," *Current Biology* 21 (2011): R994–95.

Chapter 7. How Did Reality Denial Emerge?

1   Of course, the denial mechanism might have been in use in another unknown context and then gotten co-opted for the mortality-denial purpose.

2   Derek Bickerton, *Adam's Tongue: How Humans Made Language, How Language Made Humans* (Hill and Wang, 2009).

3   Bernard Chapais, *Primeval Kinship: How Pair-Bonding Gave Birth to Human Society* (Harvard University Press, 2010).

4   Robert Trivers, *The Folly of Fools: The Logic of Deceit and Self-Deception in Human Life* (Basic Books, 2011). Ian Leslie also argues that lying is central to who we are as humans in *Born Liars: Why We Can't Live without Deceit* (Quercus, 2011).

5   Charles Darwin, *The Descent of Man, and Selection in Relation to Sex* (John Murray, 1871).

6   For examples, see OT Eldakar, DL Farrell, and DS Wilson, "Selfish Punishment: Altruism Can Be Maintained by Competition among Cheaters," *Journal of Theoretical Biology* 249 (2007): 198–205; H Ohtsuki, Y Iwasa, and MA Nowak, "Indirect Reciprocity Provides Only a Narrow Margin of Efficiency for Costly Punishment," *Nature* 457 (2009): 79–82; EG Weyl, ME Frederickson, DW Yu, and NE Pierce, "Economic Contract Theory Tests Models of Mutualism," *Proceedings of the National Academy of Sciences of the United States of America* 107 (2010): 15712–16; JH Fowler and NA Christakis, "Cooperative Behavior Cascades in Human Social Networks," *Proceedings of the National Academy of Sciences of the United States of America* 107 (2010): 5334–38; S Mathew and R Boyd, "Punishment Sustains Large-Scale Cooperation in Prestate Warfare," *Proceedings of the National Academy of Sciences of the United States of America* 108 (2011): 11375–80.

7   Simon Baron-Cohen, *The Science of Evil: On Empathy and the Origins of Cruelty* (Basic Books, 2012).

8   Steven Pinker, *The Better Angels of Our Nature: Why Violence Has Declined* (Viking, 2011).

Chapter 8. Evidence for Reality Denial Is All Around Us!

1   Stephen W. Hawking, *A Brief History of Time: From the Big Bang to Black Holes* (Bantam, 1988).

2   Sean M. Carroll, *From Eternity to Here: The Quest for the Ultimate Theory of Time* (ONEWorld, 2012).

3   Brian Greene, *The Hidden Reality: Parallel Universes and the Deep Laws of the Cosmos* (Vintage, 2011).

4   All objective evidence was contrary to these beliefs, but the US and UK governments gave them an air of respectability by constant

repetition. And there is no doubt that most of those in government who made these assertions truly believed what they were saying. After all, as we said earlier, if one can believe one's own untruth, it is much easier to put it forward in a convincing manner.

5   Robert Kurzban, *Why Everyone (Else) Is a Hypocrite: Evolution and the Modular Mind* (Princeton University Press, 2012).

6   Michael Shermer, *The Believing Brain: From Ghosts and Gods to Politics and Conspiracies—How We Construct Beliefs and Reinforce Them as Truths* (Times Books, 2011).

7   Jonathan Haidt, *The Righteous Mind: Why Good People Are Divided by Politics and Religion* (Pantheon, 2012).

8   Taken from http://www.moralfoundations.org/. The foundations are (1) care/harm, (2) fairness/cheating, (3) liberty/oppression, (4) loyalty/betrayal, (5) authority/subversion, and (6) sanctity/degradation.

9   Jack Herer, *The Emperor Wears No Clothes: The Authoritative Historical Record of Cannabis and the Conspiracy against Marijuana* (AH HA Publishing, 2010); Julie Holland, *The Pot Book: A Complete Guide to Cannabis* (Park Street Press, 2010).

10   E Manrique-Garcia, S Zammit, C Dalman, T Hemmingsson, S Andreasson, and P Allebeck, "Cannabis, Schizophrenia, and Other Non-Affective Psychoses: 35 Years of Follow-Up of a Population-Based Cohort," *Psychological Medicine* 42 (2012): 1321–28; MF Griffith-Lendering, JT Wigman, A Prince van Leeuwen, SC Huijbregts, AC Huizink, J Ormel, FC Verhulst, J van Os, H Swaab, and WA Vollebergh, "Cannabis Use and Vulnerability for Psychosis in Early Adolescence—A Trails Study," *Addiction* (2012): Epub ahead of print.

11   MH Meier, A Caspi, A Ambler, H Harrington, R Houts, RS Keefe, K McDonald, A Ward, R Poulton, and TE Moffitt, "Persistent Cannabis Users Show Neuropsychological Decline from Childhood to Midlife," *Proceedings of the National Academy of Sciences of the United States of America* 109 (2012): E2657–64.

12   I Danovitch and DA Gorelick, "State of the Art Treatments for Cannabis Dependence," *Psychiatric Clinics of North America* 35 (2012): 309–26.

13   Walter C. Willett, *Eat, Drink, and Be Healthy: The Harvard Medi-*

cal School Guide to Healthy Eating (Free Press, 2005); GA Colditz, KY Wolin, and S Gehlert, "Applying What We Know to Accelerate Cancer Prevention," *Science Translational Medicine* 4 (2012): 127rv4; C Ash, P Kiberstis, E Marshall, and J Travis, "Disease Prevention: It Takes More Than an Apple a Day," *Science* 337 (2012): 1466–67 (see also other articles in this issue).

14 World Cancer Research Fund/American Institute for Cancer Research, "Food, Nutrition, Physical Activity, and the Prevention of Cancer: A Global Perspective" (American Institute for Cancer Research, 2007). This report issued eight recommendations (and two special recommendations) on diet, physical activity, and weight management for cancer prevention. These were based on the most comprehensive collection of available evidence to date. But little has changed with regard to human behavior in relation to these risk factors.

15 Florence Williams, *Breasts: A Natural and Unnatural History* (W. W. Norton, 2013).

16 For an example, see HD Nelson, B Zakher, A Cantor, R Fu, J Griffin, ES O'Meara, DS Buist, K Kerlikowske, NT van Ravesteyn, A Trentham-Dietz, JS Mandelblatt, and DL Miglioretti, "Risk Factors for Breast Cancer for Women Aged 40 to 49 Years: A Systematic Review and Meta-Analysis," *Annals of Internal Medicine* 156 (2012): 635–48.

17 A Varki, "Addressing the Major Challenge for Women in Academia: 'It's Proximate Childcare, Stupid!'" American Society for Cell Biology newsletter (April 2008): 7–8.

18 World Heart Federation newsletter, January 2010.

19 A Pan, Q Sun, AM Bernstein, MB Schulze, JE Manson, MJ Stampfer, WC Willett, and FB Hu, "Red Meat Consumption and Mortality: Results From 2 Prospective Cohort Studies," *Archives of Internal Medicine* 172 (2012): 555–63.

20 http://www.usdebtclock.org/.

21 Daniel Quinn, *The Story of B* (Bantam, 1996).

22 Daniel Lord Smail, *On Deep History and the Brain* (University of California Press, 2008).

23 Norma Hayes Bagnall, *On Shaky Ground: The New Madrid Earthquakes of 1811–1812* (University of Missouri Press, 1996).

24   Seth Stein, *Disaster Deferred: A New View of Earthquake Hazards in the New Madrid Seismic Zone* (Columbia University Press, 2010); R Monastersky, "Seth Stein: The Quake Killer," *Nature* 479 (2011): 166–70.

25   Gina Kolata, *Flu: The Story of the Great Influenza Pandemic of 1918 and the Search for the Virus That Caused It* (Touchstone, 2001); John M. Barry, *The Great Influenza: The Story of the Deadliest Pandemic in History* (Penguin Books, 2005).

26   Walter Isaacson, *Steve Jobs* (Simon and Schuster, 2011).

27   Jobs may have tried a vegan diet, acupuncture, herbal remedies, and other treatments, and perhaps even consulted a psychic. He was also likely influenced by a clinic that advised unproven approaches such as juice fasts and bowel cleansings.

28   Richard Dawkins, *The Magic of Reality: How We Know What's Really True* (Free Press, 2011).

Chapter 9. Too Smart for Our Own Good

1   Michael Shermer, *In Darwin's Shadow: The Life and Science of Alfred Russel Wallace—A Biographical Study on the Psychology of History* (Oxford University Press, 2002).

2   Since evolution has no plan, it obviously cannot preselect for something. A better term is *exaptation*: see SJ Gould and ES Vrba, "Exaptation—a Missing Term in the Science of Form," *Paleobiology* 8 (1982): 4–15.

3   *Contributions to the Theory of Natural Selection*, 1870.

4   Bhaskar Ramamurthy, director of the Indian Institute of Technology in Madras, alerted me to this passage, cited in T. G. K. Murthy's *Swami Vivekananda: An Intuitive Scientist* (Sri Ramakrishna Math, 2012):

> In a discourse with his disciples, Vivekananda said, "In the animal kingdom we really see such laws as struggle for existence, survival of the fittest, etc., evidently at work. Therefore, Darwin's theory seems true to a certain extent. But in the human kingdom, where there is the manifestation of rationality, we find just the reverse of those laws. For instance, in those whom we consider really great

men or ideal characters, we scarcely observe any external struggle. In the animal kingdom instinct prevails; but the more a man advances, the more he manifests rationality. For this reason, progress in the rational human kingdom cannot be achieved, like that in the animal kingdom, by the destruction of others! The highest evolution of man is effected through sacrifice alone. A man is great among his fellows in proportion as he can sacrifice for the sake of others, while in the lower strata of the animal kingdom, that animal is the strongest which can kill the greatest number of animals. Hence the struggle theory is not equally applicable to both kingdoms. Man's struggle is in the mental sphere. A man is greater in proportion as he can control his mind. When the mind's activities are perfectly at rest, the Atman manifests Itself. The struggle which we observe in the animal kingdom for the preservation of the gross body has its use in the human plane of existence for gaining mastery over the mind or for attaining the state of balance. Like a living tree and its reflection in the water of a tank, we find opposite kinds of struggle in the animal and human kingdoms."

The original of this passage appears in *The Complete Works of Swami Vivekananda*, volume 7 (Vedanta Press & Bookshop, 1972).

5   RM French, "The Turing Test: The First 50 Years," *Trends in Cognitive Sciences* 4 (2000): 115–22.

6   http://longbets.org/1/#terms.

7   *Contributions to the Theory of Natural Selection*, 1870.

8   Andrew Keen, *Digital Vertigo: How Today's Online Social Revolution Is Dividing, Diminishing, and Disorienting Us* (St. Martin's Griffin, 2013); Nicholas Carr, *The Shallows: What the Internet Is Doing to Our Brains* (W. W. Norton, 2011); Sherry Turkle, *Alone Together: Why We Expect More from Technology and Less from Each Other* (Basic Books, 2011).

9   Charles Seife, *Zero: The Biography of a Dangerous Idea* (Penguin Books, 2000).

10   Tobias Dantzig, *Number: The Language of Science*, 4th ed. (Macmillan, 1954).

11   Seife, *Zero*.

12   Linda Stone and Paul F. Lurquin, with L. Luca Cavalli-Sforza, *Genes, Culture, and Human Evolution: A Synthesis* (Blackwell, 2007).

13 Peter J. Richerson and Robert Boyd, *Not by Genes Alone: How Culture Transformed Human Evolution* (University of Chicago Press, 2005).

14 Richard Dawkins, *The Selfish Gene*, 30th anniversary ed. (Oxford University Press, 2006).

15 Alex Mesoudi, *Cultural Evolution: How Darwinian Theory Can Explain Human Culture and Synthesize the Social Sciences* (University of Chicago Press, 2011).

16 RG Roberts and MI Bird, "Evolutionary Anthropology: Homo 'Incendius,'" *Nature* 485 (2012): 586–87.

17 Bruce H. Weber and David J. Depew, eds., *Evolution and Learning: The Baldwin Effect Reconsidered* (MIT Press, 2003); Terrence Deacon, *The Symbolic Species: The Co-Evolution of Language and the Brain* (W.W. Norton, 1997).

18 F. John Odling-Smee, Kevin N. Laland, and Marcus W. Feldman, *Niche Construction: The Neglected Process in Evolution* (Princeton University Press, 2003).

19 J Henrich, "Demography and Cultural Evolution: How Adaptive Cultural Processes Can Produce Maladaptive Losses—The Tasmanian Case," *American Antiquity* 69 (2004): 197–214.

20 Iain Davidson, "Tasmanian Aborigines and the Origins of Language," in John Mulvaney and Hugh Tyndale-Biscoe, eds., *Rediscovering Recherche Bay* (Academy of the Social Sciences in Australia, 2007): 69–85.

21 After writing this I found a similar idea in David Barash's book *Natural Selections: Selfish Altruists, Honest Liars, and Other Realities of Evolution* (Bellevue Literary Press, 2012): "Imagine that you could exchange a newborn baby from the mid-Pleistocene—say, 50,000 years ago—with a 21st-century newborn. Both children—the one fast-forwarded no less than the other brought back in time—would doubtless grow up to be normal members of their society, essentially indistinguishable from their colleagues who had been naturally born into it . . . . But switch a modern human adult and adults from the late Ice Age, and there would be Big Trouble, either way."

22 A Senghas, S Kita, and A Ozyurek, "Children Creating Core Properties of Language: Evidence from an Emerging Sign Language in Nicaragua," *Science* 305 (2004): 1779–82.

23   W Sandler, M Aronoff, I Meir, and C Padden, "The Gradual Emergence of Phonological Form in a New Language," *Natural Language and Linguistic Theory* 29 (2011): 503–43.

24   Douglas K. Candland, *Feral Children and Clever Animals: Reflections on Human Nature* (Oxford University Press, 1995); Michael Newton, *Savage Girls and Wild Boys: A History of Feral Children*, reprint ed. (Picador, 2004); Adriana S. Benzaquen, *Encounters with Wild Children: Temptation and Disappointment in the Study of Human Nature* (McGill–Queens University Press, 2006).

25   Russ Rymer, *Genie: A Scientific Tragedy* (HarperPerennial, 1994).

26   RN Campbell and R Grieve, "Royal Investigations of the Origin of Language," *Historiographia linguistica* 9 (1982): 1–2.

27   Abul Fazl, *The Akbarnama of Abu-L-Fazl*, vol. 3, translated by Henry Beveridge, reprint ed. (Low Price Publications, 1993): 581–582.

28   Henry Beveridge, "Father Jerome Xavier," *Journal of the Asiatic Society of Bengal* 57, part 1 (1888): 33–39.

29   Sir Henry M. Elliot, *The History of India, as Told by Its Own Historians: The Muhammadan Period*, ed. by John Dowson, reprint ed. (Islamic Book Service, 1979).

30   A Varki, DH Geschwind, and EE Eichler, "Explaining Human Uniqueness: Genome Interactions with Environment, Behaviour, and Culture," *Nature Reviews Genetics* 9 (2008): 749–63. The second part of this review contains many ideas relevant to this book.

31   Sarah Blaffer Hrdy, *Mother Nature: A History of Mothers, Infants, and Natural Selection* (Pantheon, 1999).

32   AC Hannah and B Brotman, "Procedures for Improving Maternal Behavior in Captive Chimpanzees," *Zoo Biology* 9 (1990): 233–40.

33   Richard Louv, *Last Child in the Woods: Saving Our Children from Nature-Deficit Disorder* (Algonquin Books, 2008).

34   Alvin Toffler, *Future Shock* (Random House, 1970).

35   NG Blurton Jones, K Hawkes, and JF O'Connell, "Antiquity of Postreproductive Life: Are There Modern Impacts on Hunter-Gatherer Postreproductive Life Spans?" *American Journal of Human Biology* 14, no. 2 (2002): 184–205.

36   AS Fauci and FS Collins, "Benefits and Risks of Influenza Research: Lessons Learned," *Science* 336 (2012): 1522–23.

37   Stacy Mintzer Herlihy and E. Allison Hagood, *Your Baby's Best Shot: Why Vaccines Are Safe and Save Lives* (Rowman & Littlefield, 2012); Seth Mnookin, *The Panic Virus: The True Story Behind the Vaccine-Autism Controversy* (Simon and Schuster, 2012).

Chapter 10. A Tale of Two Futures: Are You a Pessimist or an Optimist?

1   http://ucsdnews.ucsd.edu/features/global_warming_solutions_dependent_on_oneness_of_humanity_dalai_lama_tells/.

2   Paul Gilding, *The Great Disruption: Why the Climate Crisis Will Bring On the End of Shopping and the Birth of a New World* (Bloomsbury, 2012).

3   Al Gore, *An Inconvenient Truth: The Crisis of Global Warming* (Viking, 2007).

4   Fred Guterl, *The Fate of the Species: Why the Human Race May Cause Its Own Extinction and How We Can Stop It* (Bloomsbury, 2012).

5   Michael E. Mann, *The Hockey Stick and the Climate Wars: Dispatches from the Front Lines* (Columbia University Press, 2012).

6   PJ Crutzen, "Anthropocene Man," *Nature* 467 (2010): S10.

7   S Rahmstorf and D Coumou, "Increase of Extreme Events in a Warming World," *Proceedings of the National Academy of Sciences of the United States of America* 108 (2011): 17905–9; D Coumou and S Rahmstorf, "A Decade of Weather Extremes," *Nature Climate Change* (2012), DOI: 10.1038/Nclimate1452.

8   RA Kerr, "Climate Change: Ice-Free Arctic Sea May Be Years, Not Decades, Away," *Science* 337 (2012): 1591.

9   S Barker, G Knorr, RL Edwards, F Parrenin, AE Putnam, LC Skinner, E Wolff, and M Ziegler, "800,000 Years of Abrupt Climate Variability," *Science* 334 (2011): 347–51.

10   William F. Ruddiman, *Plows, Plagues, and Petroleum: How Humans Took Control of Climate* (Princeton University Press, 2010).

11   J Hansen, M Sato, and R Ruedy, "Perception of Climate Change," *Proceedings of the National Academy of Sciences of the United States of America* 109 (2012): E2415–23.

12 http://www.unep.org/newscentre/default.aspx?DocumentID=2700 &ArticleID=9353.

13 http://www.resource.uk.com/article/UK/UN_Climate_Change_ Conference_extends_Kyoto_Protocol-2536.

14 http://www.resource.uk.com/article/UK/UN_Climate_Change_ Conference_extends_Kyoto_Protocol-2536.

15 http://www.ncdc.noaa.gov/sotc/national/2012/13.

16 http://www.ncdc.noaa.gov/sotc/.

17 Rachel Carson, *Silent Spring*, 40th anniversary ed. (Houghton Mifflin, 2002).

18 Paul R. Ehrlich, *The Population Bomb* (Sierra Club, 1969).

19 PR Ehrlich and AH Ehrlich, "The Population Bomb Revisited," *The Electronic Journal of Sustainable Development* 1 (2009): 63–71.

20 J Shepherd, D Iglesias-Rodriguez, and A Yool, "Geo-Engineering Might Cause, Not Cure, Problems," *Nature* 449 (2007): 781; PW Boyd, "Ranking Geo-Engineering Schemes," *Nature Geoscience* 1 (2008): 722–24; JJ Blackstock and JCS Long, "The Politics of Geo-engineering," *Science* 327 (2010): 527–29; A Robock, M Bunzl, B Kravitz, and GL Stenchikov, "A Test for Geoengineering?" *Science* 327 (2010): 530–31.

21 John Broome, *Climate Matters: Ethics in a Warming World* (W. W. Norton, 2012).

22 DJ Zaelke and V Ramanathan, "Going beyond Carbon Dioxide," *New York Times*, Opinion pages, December 6, 2012.

23 http://www.carbonnationmovie.com/.

24 Barbara Kingsolver, *Flight Behavior* (HarperCollins, 2012).

25 Anders Wijkman and Johan Rockström, *Bankrupting Nature: Denying Our Planetary Boundaries* (Routledge, 2012).

26 J Ratcliffe, "Social Justice and the Demographic Transition: Lessons from India's Kerala State," *International Journal of Health Services* 8 (1978): 123–144.

27 Stephen Webb, *If the Universe Is Teeming with Aliens... Where Is Everybody? Fifty Solutions to the Fermi Paradox and the Problem of Extraterrestrial Life* (Springer, 2010).

28 Frank Drake and Dava Sobel, *Is Anyone Out There? The Scientific Search for Extraterrestrial Intelligence* (Delta, 1994).

29   In the *Star Trek* series, the Prime Directive is Starfleet's General Order 1, the most prominent guiding principle.

30   Robert Trivers, *The Folly of Fools: The Logic of Deceit and Self-Deception in Human Life* (Basic Books, 2011).

31   American Academy of Arts and Sciences and American Philosophical Society, *The Public Good: Knowledge as the Foundation for a Democratic Society* (American Academy of Arts and Sciences, 2008).

32   Michael Specter, *Denialism: How Irrational Thinking Hinders Scientific Progress, Harms the Planet, and Threatens Our Lives* (Penguin Press, 2009).

Chapter 11. On the Positive Value of Human Reality Denial

1   R Schulz, J Bookwala, JE Knapp, M Scheier, and GM Williamson, "Pessimism, Age, and Cancer Mortality," *Psychology and Aging* 11 (1996): 304.

2   MM Kogon, A Biswas, D Pearl, RW Carlson, and D Spiegel, "Effects of Medical and Psychotherapeutic Treatment on the Survival of Women with Metastatic Breast Carcinoma," *Cancer* 80 (1997): 225–30.

3   PJ Goodwin, M Leszcz, M Ennis, J Koopmans, L Vincent, H Guther, E Drysdale, M Hundleby, HM Chochinov, M Navarro, M Speca, and J Hunter, "The Effect of Group Psychosocial Support on Survival in Metastatic Breast Cancer," *New England Journal of Medicine* 345 (2001): 1719–26.

4   JC Weeks, PJ Catalano, A Cronin, MD Finkelman, JW Mack, NL Keating, and D Schrag, "Patients' Expectations about Effects of Chemotherapy for Advanced Cancer," *New England Journal of Medicine* 367 (2012): 1616–25.

5   TJ Smith and DL Longo, "Talking with Patients about Dying," *New England Journal of Medicine* 367 (2012): 1651–52.

6   T Sharot, AM Riccardi, CM Raio, and EA Phelps, "Neural Mechanisms Mediating Optimism Bias," *Nature* 450, no. 7166 (2007): 102–5; T Sharot, CW Korn, and RJ Dolan, "How Unrealistic Optimism Is Maintained in the Face of Reality," *Nature Neuroscience*

14 (2011): 1475–79; T Sharot, R Kanai, D Marston, CW Korn, G Rees, and RJ Dolan, "Selectively Altering Belief Formation in the Human Brain," *Proceedings of the National Academy of Sciences of the United States of America* 109 (2012): 17058–62.

7   SL Keng, MJ Smoski, and CJ Robins, "Effects of Mindfulness on Psychological Health: A Review of Empirical Studies," *Clinical Psychology Review* 31 (2011): 1041–56.

8   Krish Sathian of Emory University pointed out to me that this is actually true at some level, since what we experience is really a neural construct that in many cases deviates from reality, as exemplified by a variety of illusions, from the optical (e.g., Necker cubes) to the kinesthetic (e.g., phantom limbs).

9   Apparently this term was based on "The Menagerie" a two-part episode of the classic *Star Trek* series.

10   Christopher Hitchens, *Mortality* (Twelve, 2012).

11   http://www.deathcafe.com/. The objective of a Death Cafe is "to increase awareness of death with a view to helping people make the most of their (finite) lives."

Chapter 12. Explaining the Mysterious Origin of Us

1   Bernard Wood, *Human Evolution: A Brief Insight* (Sterling, 2011).

2   Robert Boyd and Joan B. Silk, *How Humans Evolved* (W. W. Norton, 2011).

3   Chris Stringer, *Lone Survivors: How We Came to Be the Only Humans on Earth* (Henry Holt, 2012).

4   Ian Tattersall, *Masters of the Planet: The Search for Our Human Origins* (Palgrave Macmillan, 2012).

5   Camilo J. Cela-Conde and Francisco J. Ayala, *Human Evolution: Trails from the Past* (Oxford University Press, 2007).

6   I was fortunate to speak at the European Society for the Study of Human Evolution meeting in September of 2012, where I heard some of this new information. But by the time this book comes out in 2013, there will likely be even more relevant data. Also, there are many details of the story I will gloss over somewhat in an effort to focus on the main point.

7   M Meyer, M Kircher, MT Gansauge, H Li, F Racimo, S Mallick, JG Schraiber, F Jay, K Prufer, C de Filippo, PH Sudmant, C Alkan, Q Fu, R Do, N Rohland, A Tandon, M Siebauer, RE Green, K Bryc, AW Briggs, U Stenzel, J Dabney, J Shendure, J Kitzman, MF Hammer, MV Shunkov, AP Derevianko, N Patterson, AM Andres, EE Eichler, M Slatkin, D Reich, J Kelso, and S Pääbo, "A High-Coverage Genome Sequence from an Archaic Denisovan Individual," *Science* 338 (2012): 222–26.

8   A Scally and R Durbin, "Revising the Human Mutation Rate: Implications for Understanding Human Evolution," *Nature Reviews: Genetics* 13 (2012): 745–53; E Callaway, "Studies Slow the Human DNA Clock," *Nature* 489 (2012): 343–44; A Gibbons, "Turning Back the Clock: Slowing the Pace of Prehistory," *Science* 338 (2012): 189–91.

9   S McBrearty and AS Brooks, "The Revolution That Wasn't: A New Interpretation of the Origin of Modern Human Behavior," *Journal of Human Evolution* 39 (2000): 453–563.

10   S McBrearty, "Palaeoanthropology: Sharpening the Mind," *Nature* 491 (2012): 531–32.

11   TD Weaver, "Did a Discrete Event 200,000–100,000 Years Ago Produce Modern Humans?" *Journal of Human Evolution* 63 (2012): 121–26.

12   O Bar-Yosef, K Boyle, P Mellars, and C Stringer, eds., *Rethinking the Human Revolution: New Behavioural and Biological Perspectives on the Origin and Dispersal of Modern Humans* (McDonald Institute for Archaeological Research, 2007).

13   Chris Stringer, *Lone Survivors: How We Came to Be the Only Humans on Earth* (Henry Holt, 2012).

14   Ian Tattersall, *Masters of the Planet: The Search for Our Human Origins* (Palgrave Macmillan, 2012).

15   Colin Renfrew, *Prehistory: The Making of the Human Mind* (Modern Library, 2009).

16   William Calvin, *The Ascent of Mind: Ice Age Climates and the Evolution of Intelligence* (iUniverse, 2001).

17   A Robock, CM Ammann, L Oman, D Shindell, S Levis, and G Stenchikov, "Did the Toba Volcanic Eruption of ~74k BP Produce Widespread Glaciation?" *Journal of Geophysical Research*

114 (2009): D10107; M. Williams, "Did the 73 ka Toba Super-Eruption Have an Enduring Effect? Insights from Genetics, Prehistoric Archaeology, Pollen Analysis, Stable Isotope Geochemistry, Geomorphology, Ice Cores, and Climate Models," *Quaternary International* 269 (2011): 87–93.

18   SH Ambrose, "Late Pleistocene Human Population Bottlenecks, Volcanic Winter, and Differentiation of Modern Humans," *Journal of Human Evolution* 34 (1998): 623–51.

19   X Wang, N Mitra, I Secundino, K Banda, P Cruz, V Padler-Karavani, A Verhagen, C Reid, M Lari, E Rizzi, C Balsamo, G Corti, G De Bellis, L Longo, W Beggs, D Caramelli, SA Tishkoff, T Hayakawa, ED Green, JC Mullikin, V Nizet, J Bui, and A Varki, "Specific Inactivation of Two Immunomodulatory Siglec Genes during Human Evolution," *Proceedings of the National Academy of Sciences of the United States of America* 109 (2012): 9935–40.

20   MF Hammer, AE Woerner, FL Mendez, JC Watkins, and JD Wall, "Genetic Evidence for Archaic Admixture in Africa," *Proceedings of the National Academy of Sciences of the United States of America* 108 (2011): 15123–28; J Lachance, B Vernot, CC Elbers, B Ferwerda, A Froment, JM Bodo, G Lema, W Fu, TB Nyambo, TR Rebbeck, K Zhang, JM Akey, and SA Tishkoff, "Evolutionary History and Adaptation from High-Coverage Whole-Genome Sequences of Diverse African Hunter-Gatherers," *Cell* 150 (2012): 457–69.

21   L Abi-Rached, MJ Jobin, S Kulkarni, A McWhinnie, K Dalva, L Gragert, F Babrzadeh, B Gharizadeh, M Luo, FA Plummer, J Kimani, M Carrington, D Middleton, R Rajalingam, M Beksac, SG Marsh, M Maiers, LA Guethlein, S Tavoularis, AM Little, RE Green, PJ Norman, and P Parham, "The Shaping of Modern Human Immune Systems by Multiregional Admixture with Archaic Humans," *Science* 334 (2011): 89–94; FL Mendez, JC Watkins, and MF Hammer, "Global Genetic Variation at OAS1 Provides Evidence of Archaic Admixture in Melanesian Populations," *Molecular Biology and Evolution* 29 (2012): 1513–20; FL Mendez, JC Watkins, and MF Hammer, "A Haplotype at STAT2 Introgressed from Neanderthals and Serves as a Candidate of Positive Selection in Papua New Guinea," *American Journal of Human Genetics* 91 (2012): 265–74.

22   CM Schlebusch, P Skoglund, P Sjodin, LM Gattepaille, D Hernandez, F Jay, S Li, M De Jongh, A Singleton, MG Blum, H Soodyall, and M Jakobsson, "Genomic Variation in Seven Khoe-San Groups Reveals Adaptation and Complex African History," *Science* 338, no. 6105 (2012): 374–79.

23   L Abi-Rached, et al., "The Shaping of Modern Human Immune Systems," 89–94; FL Mendez, et al., "Global Genetic Variation," 1513–20; FL Mendez, JC Watkins, and MF Hammer, "A Haplotype at STAT2," 265–74.

24   T Higham, T Compton, C Stringer, R Jacobi, B Shapiro, E Trinkaus, B Chandler, F Groning, C Collins, S Hillson, P O'Higgins, C FitzGerald, and M Fagan, "The Earliest Evidence for Anatomically Modern Humans in Northwestern Europe," *Nature* 479 (2011): 521–24, and references therein.

25   JJ Hublin, F Spoor, M Braun, F Zonneveld, and S Condemi, "A Late Neanderthal Associated with Upper Palaeolithic Artefacts," *Nature* 381 (1996): 224–26; JJ Hublin, S Talamo, M Julien, F David, N Connet, P Bodu, B Vandermeersch, and MP Richards, "Radiocarbon Dates from the Grotte du Renne and Saint-Cesaire Support a Neandertal Origin for the Chatelperronian," *Proceedings of the National Academy of Sciences of the United States of America* 109 (2012): 18743–48; S Benazzi, K Douka, C Fornai, CC Bauer, O Kullmer, J Svoboda, I Pap, F Mallegni, P Bayle, M Coquerelle, S Condemi, A Ronchitelli, K Harvati, and GW Weber, "Early Dispersal of Modern Humans in Europe and Implications for Neanderthal Behaviour," *Nature* 479 (2011): 525–28.

26   JF Hoffecker, "Out of Africa: The Spread of Modern Humans in Europe," *Proceedings of the National Academy of Sciences of the United States of America* 106 (2009): 16040–45.

27   JJ Hublin, "The Earliest Modern Human Colonization of Europe," *Proceedings of the National Academy of Sciences of the United States of America* 109 (2012): 13471–72.

28   B Roozendaal, BS McEwen, and S Chattarji, "Stress, Memory, and the Amygdala," *Nature Reviews: Neuroscience* 10 (2009): 423–33; JP Johansen, CK Cain, LE Ostroff, and JE LeDoux, "Molecular Mechanisms of Fear Learning and Memory," *Cell* 147 (2011): 509–24; RJ Davidson and BS McEwen, "Social Influences on Neu-

roplasticity: Stress and Interventions to Promote Well-Being," *Nature Neuroscience* 15 (2012): 689–95.

29   CN Carlo, L Stefanacci, K Semendeferi, and CF Stevens, "Comparative Analyses of the Neuron Numbers and Volumes of the Amygdaloid Complex in Old and New World Primates," *Journal of Comparative Neurology* 518 (2010): 1176–98; N Barger, L Stefanacci, CM Schumann, CC Sherwood, J Annese, JM Allman, JA Buckwalter, PR Hof, and K Semendeferi, "Neuronal Populations in the Basolateral Nuclei of the Amygdala Are Differentially Increased in Humans Compared with Apes: A Stereological Study," *Journal of Comparative Neurology* 520 (2012): 3035–54; JK Rilling, J Scholz, TM Preuss, MF Glasser, BK Errangi, and TE Behrens, "Differences between Chimpanzees and Bonobos in Neural Systems Supporting Social Cognition," *Social Cognitive and Affective Neuroscience* 7 (2012): 369–79.

30   C Ecker, J Suckling, SC Deoni, MV Lombardo, ET Bullmore, S Baron-Cohen, M Catani, P Jezzard, A Barnes, AJ Bailey, SC Williams, and DG Murphy, "Brain Anatomy and Its Relationship to Behavior in Adults with Autism Spectrum Disorder: A Multicenter Magnetic Resonance Imaging Study," *Archives of General Psychiatry* 69 (2012): 195–209; D Kliemann, I Dziobek, A Hatri, J Baudewig, and HR Heekeren, "The Role of the Amygdala in Atypical Gaze on Emotional Faces in Autism Spectrum Disorders," *Journal of Neuroscience* 32 (2012): 9469–76; CW Nordahl, R Scholz, X Yang, MH Buonocore, T Simon, S Rogers, and DG Amaral, "Increased Rate of Amygdala Growth in Children Aged 2 to 4 Years with Autism Spectrum Disorders: A Longitudinal Study," *Archives of General Psychiatry* 69 (2012): 53–61.

31   BW Haas, F Hoeft, YM Searcy, D Mills, U Bellugi, and A Reiss, "Individual Differences in Social Behavior Predict Amygdala Response to Fearful Facial Expressions in Williams Syndrome," *Neuropsychologia* 48 (2010): 1283–88; M Jabbi, JS Kippenhan, P Kohn, S Marenco, CB Mervis, CA Morris, A Meyer-Lindenberg, and KF Berman, "The Williams Syndrome Chromosome 7q11.23 Hemideletion Confers Hypersocial, Anxious Personality Coupled with Altered Insula Structure and Function," *Proceedings of the*

National Academy of Sciences of the United States of America 109 (2012): E860–66.

Chapter 13. Future Directions

1   JH Langdon, "Umbrella Hypotheses and Parsimony in Human Evolution: A Critique of the Aquatic Ape Hypothesis," *Journal of Human Evolution* 33 (1997): 479–94.

2   Karl R. Popper, *The Logic of Scientific Discovery,* revised ed. (Unwin Hyman, 1980).

3   Thomas S. Kuhn, *The Structure of Scientific Revolutions*, 50th anniversary ed. (University of Chicago Press, 2012).

4   M Bedny, A Pascual-Leone, and RR Saxe, "Growing Up Blind Does Not Change the Neural Bases of Theory of Mind," *Proceedings of the National Academy of Sciences* 106 (2009): 11312–17; JP Mitchell, "Inferences about Mental States," *Philosophical Transactions of the Royal Society B: Biological Sciences* 364 (2009): 1309–16.; DI Tamir and JP Mitchell, "Neural Correlates of Anchoring-and-Adjustment during Mentalizing," *Proceedings of the National Academy of Sciences of the United States of America* 107 (2010): 10827–32; DI Tamir and JP Mitchell, "Anchoring and Adjustment during Social Inferences," *Journal of Experimental Psychology: General* (2012), DOI: 10.1037/a0028232; L Young, D Dodell-Feder, and R Saxe, "What Gets the Attention of the Temporo-Parietal Junction? An fMRI Investigation of Attention and Theory of Mind," *Neuropsychologia* 48 (2010): 2658–64.

5   J Klackl, E Jonas, and M Kronbichler, "Existential Neuroscience: Neurophysiological Correlates of Proximal Defenses against Death-Related Thoughts," *Social Cognitive and Affective Neuroscience* (2012): Epub ahead of print; M Quirin, A Loktyushin, J Arndt, E Kustermann, YY Lo, J Kuhl, and L Eggert, "Existential Neuroscience: A Functional Magnetic Resonance Imaging Investigation of Neural Responses to Reminders of One's Mortality," *Social Cognitive and Affective Neuroscience* 7 (2012): 193–98.

6   B Roozendaal, BS McEwen, and S Chattarji, "Stress, Memory, and the Amygdala," *Nature Reviews: Neuroscience* 10 (2009): 423–33;

JP Johansen, CK Cain, LE Ostroff, and JE LeDoux, "Molecular Mechanisms of Fear Learning and Memory," *Cell* 147 (2011): 509–24; RJ Davidson and BS McEwen, "Social Influences on Neuroplasticity: Stress and Interventions to Promote Well-Being," *Nature Neuroscience* 15 (2012): 689–95.

7   CN Carlo, L Stefanacci, K Semendeferi, and CF Stevens, "Comparative Analyses of the Neuron Numbers and Volumes of the Amygdaloid Complex in Old and New World Primates," *Journal of Comparative Neurology* 518 (2010): 1176–98; N Barger, L Stefanacci, CM Schumann, CC Sherwood, J Annese, JM Allman, JA Buckwalter, PR Hof, and K Semendeferi, "Neuronal Populations in the Basolateral Nuclei of the Amygdala Are Differentially Increased in Humans Compared with Apes: A Stereological Study," *Journal of Comparative Neurology* 520 (2012): 3035–54; JK Rilling, J Scholz, TM Preuss, MF Glasser, BK Errangi, and TE Behrens, "Differences between Chimpanzees and Bonobos in Neural Systems Supporting Social Cognition," *Social Cognitive and Affective Neuroscience* 7 (2012): 369–79.

8   C Ecker, J Suckling, SC Deoni, MV Lombardo, ET Bullmore, S Baron-Cohen, M Catani, P Jezzard, A Barnes, AJ Bailey, SC Williams, and DG Murphy, "Brain Anatomy and Its Relationship to Behavior in Adults with Autism Spectrum Disorder: A Multicenter Magnetic Resonance Imaging Study," *Archives of General Psychiatry* 69 (2012): 195–209; D Kliemann, I Dziobek, A Hatri, J Baudewig, and HR Heekeren, "The Role of the Amygdala in Atypical Gaze on Emotional Faces in Autism Spectrum Disorders," *Journal of Neuroscience* 32 (2012): 9469–76; CW Nordahl, R Scholz, X Yang, MH Buonocore, T Simon, S Rogers, and DG Amaral, "Increased Rate of Amygdala Growth in Children Aged 2 to 4 Years with Autism Spectrum Disorders: A Longitudinal Study," *Archives of General Psychiatry* 69 (2012): 53–61.

Epilogue

1   I have borrowed this phrase from Charles Darwin, who used it to summarize his approach to a novel theory (descent by evolution

via natural selection) that seemed to explain everything he knew but that he could not prove by any experiment. The phrase is also within the title of a book by the great evolutionary biologist Ernst Mayr: *One Long Argument: Charles Darwin and the Genesis of Modern Evolutionary Thought* (Harvard University Press, 1993).

2   These were the words of Darwin's friend Thomas Huxley. A good idea in science will eventually be pursued to the point of either being overturned or exalted to the status of "fact." So, as eventually happened with Darwin, we hope that other scientists and thinkers will come up with ways to more precisely test the theory we have presented.

3   Michael Shermer, *The Believing Brain: From Ghosts and Gods to Politics and Conspiracies—How We Construct Beliefs and Reinforce Them as Truths* (Times Books, 2011).

4   Chris Frith, *Making Up the Mind: How the Brain Creates Our Mental World* (Blackwell, 2007).

5   Stephen Hawking and Leonard Mlodinow, *The Grand Design* (Bantam, 2012).

INDEX

ABOUT THE AUTHORS

Ajit Varki is a physician-scientist who is currently distinguished professor of medicine and cellular and molecular medicine at the University of California, San Diego (UCSD). He is also the co-director of the UCSD's Glycobiology Research and Training Center and the co-director of the Center for Academic Research and Training in Anthropogeny, affiliated with UC San Diego and the Salk Institute for Biological Studies.

Danny Brower, a cell biologist and insect geneticist, was professor and chair of molecular and cellular biology at the University of Arizona in Tucson. He died in 2007.

ABOUT TWELVE
Mission Statement

TWELVE

TWELVE was established in August 2005 with the objective of publishing no more than one book per month. We strive to publish the singular book, by authors who have a unique perspective and compelling authority. Works that explain our culture; that illuminate, inspire, provoke, and entertain. We seek to establish communities of conversation surrounding our books. Talented authors deserve attention not only from publishers but from readers as well. To sell the book is only the beginning of our mission. To build avid audiences of readers who are enriched by these works—that is our ultimate purpose.

For more information about forthcoming TWELVE books, you can visit us at www.twelvebooks.com.